Risk, Governance and Society

Volume 17

Editors
Jeryl L. Mumpower; Texas A&M University, TAMU College Station, TX, USA
Ortwin Renn, University of Stuttgart, Interdisciplinary Research Unit on Risk Governance and Sustainable Technology Development (ZIRN, Germany)

For further volumes:
http://www.springer.com/series/6643

Tom Ritchey

Wicked Problems – Social Messes

Decision Support Modelling with Morphological Analysis

Springer

Tom Ritchey, Ph.D
Ritchey Consulting AB
Swedish Morphological Society
Lyckselevägen 35
162 67 Vällingby
Sweden
ritchey@swemorph.com

ISBN 978-3-642-27076-5 ISBN 978-3-642-19653-9 (eBook)
DOI 10.1007/978-3-642-19653-9
Springer Heidelberg Dordrecht London New York

Softcover reprint of the hardcover 1st edition 2011

Cover design: eStudio Calamar S.L.

Printed on acid-free paper

Springer is part of Springer Science+Business Media (www.springer.com)

Contents

Chapter 1
Introduction

If you work with long-term social, commercial or organisational planning – or any type of policy planning that impacts *people* – then you've got *wicked problems*. You may not call them by this name, but you know what they are. They are those complex, ever changing societal and organisational planning problems that are difficult to define and structure properly because they won't keep still. They're messy, ambiguous and *reactive*, i.e. they fight back when you try to do something with them.

The term "wicked problem" was coined by Horst Rittel (Rittel, 1972; Rittel & Webber, 1973), the brilliant design theorist based at the University of Berkley (see Chap. 3). At first glance, it is not self-evident what Rittel meant by this term. Both the words "wicked" and "problem" need to be qualified: Problems are "wicked" not in the sense of being "evil", but in that they are *seriously devious* and are notoriously susceptible to the so-called "law of unintended consequences". Furthermore, as a decision maker, whatever decision you make, a good portion of the stakeholders involved are going to want your head on a block!

Also, wicked problems are not actually "problems" in the sense of having well defined and stable *problem statements*: they haven't come that far yet. This is why they have also been called *social messes* and *unstructured reality* (Ackoff, 1974; Horn 2001).

For 20 years, I worked with *wicked problems* at the Swedish Defence Research Agency (*Totalförsvarets Forskningsinstitut*, FOI) in Stockholm. Generally, these were problems of long-term defence policy, civil preparedness planning and disaster mitigation. More specifically they were about antagonistic threat scenarios, mass murder, political corruption, nuclear sabotage, failed states, uncontrolled migration and any number of distressing things that can happen to a country or population. And all of this was seen as taking place under what is called *genuine uncertainty* – i.e. there is no way to calculate the probability of something happening, and for the most part we are not even sure what might happen.

Our scenario and strategy groups would systematically look for various ways for an "aggressor" to release radioactive material, bomb public places, gas or incinerate thousands of people at a football match, sabotage the electricity grid in the middle of the Swedish winter (it almost happened), or for society to be decimated by the

T. Ritchey, *Wicked Problems – Social Messes*, Risk, Governance and Society 17,
DOI 10.1007/978-3-642-19653-9_1, © Springer-Verlag Berlin Heidelberg 2011

release of a super-virulent man-made virus. (Literally as I write this, in May of 2010, the Swedish Evening News reports that Craig Venter and friends have just constructed a synthetic organism – *Mycoplasma mycoides* JVCI-synI.0).

You tend to get callous. We would sit around laughing our heads off at all the weird and evil stuff that we could imagine being perpetrated on society – in order to think about how we might defend ourselves against it. But it was mostly a back-office curiosity: during the cold war period, the people actually running the country were not particularly interested in this sort of thing. At that time, the principle Swedish Defence Establishment concern was being *invaded* – by you know who.

Then everything changed. The Cold War ended (abruptly, in historical terms) and by the middle of the 1990s the project I was running, in order to develop new types of planning methods for what were euphemistically called *extraordinary events*, suddenly began to generate interest. In short, the end of the Cold War literally *caused* the development of computer-aided *General Morphological Analysis*.

What happened?

During the Cold War, one of the main tasks of the Division of Defence Analysis at FOI was to monitor the Soviet Union in an attempt to count, and keep track of, troops, tanks, aircraft and whatnot, in order to determine how long "we" (i.e. Swedish society and the Swedish Total Defence System) could hold out if Sweden were invaded as part of a conventional east-west war. Needless to say, this was not an exact science, and we usually felt that we could, depending on the ferocity of the invasion, make a go at it for a number of days or number of weeks – before we had to ask "somebody" (we were not supposed to use the N-word) to come to our aid. Of course, we were officially "neutral", but the history of Swedish neutrality is full of Jesuit logic. In any event, an *invaded* neutral country is, by definition, not neutral any longer.

This was the general state of affairs within the Swedish Defence establishment during the 1960s, 1970s and 1980s, until the deterioration and final break-up of the Soviet Union in the early 1990s. And at this point, the Swedish military establishment went into a sort of crisis.

In order to understand this, you have to understand that Russia – disguised as the Soviet Union after 1919 – has always been a bogeyman for Sweden. During the past 400 years (with a couple of notable parentheses) we have (rightly or wrongly) feared the Russian Bear above everything else. During the cold war period, practically all of (neutral) Sweden's national defence preparedness pointed in one single direction: east.

And the Swedish Defence was formidable. After WW2, Sweden had a *per capita* defence budget that rivalled any nation; during the 1960s and 1970s we are said to have had the world's fourth largest military air force; and (again, *per capita*) developed one of the world's largest defence export industries. We had our own fighter aircraft production, submarine production and world class artillery technology. In case of an invasion, we planned to mobilise one tenth of the population.

However, by the middle of 1990s, every intelligence service in the western world was telling us that Russia didn't have the wherewithal to invade Liechtenstein.

This was serious. We had essentially lost our enemy. For a National Defence Establishment to lose its (only) enemy is a terrible thing. Because if you lose your enemy, you are going to lose your budget!

It didn't take long before one started hearing voices: "We do not believe that the end of the cold war signifies a reduced threat spectrum. The dissolution of the Soviet Union is like taking the lid of some frightful Pandora's Box, and terrible things are going to ooze out of it in the coming 10–20 years. The threat spectrum is not going to decrease – it is going to broaden, and things are going to be more confusing and uncertain than ever before".

A brilliant defence! And, as it has turned out, essentially correct. Furthermore, it resulted in the establishment of a full-time research program: "Develop new, practical, computer based methods and instruments for long-term planning for an increasingly uncertain international situation".

During a series of diagnostic interviews with defence planners, it became clear that the defence planning system needed modelling methods and virtual laboratories that would give them enhanced power to formulate, collate, compare, test and manage hundreds or thousands of (1) possible international developments, (2) flexible strategies for such developments and, (3) flexible organisational structures to facilitate these strategies.

Such a virtual modelling environment should be able to *formulate* and *inter-relate* such diverse issues as threat assessments, technology development, national political directives, organisational structure, educational requirements, public perceptions, ethical issues, and so on and so forth. These are issues that usually cannot be (meaningfully) quantified; they contain irreducible uncertainties; they are strongly stakeholder dependent; and – politically – they are highly sensitive. This is exactly what *wicked problems* are all about.

Suddenly, my being a social anthropologist with a background in mathematics and computer science made some sort of sense. Social anthropology and sociology work with a simple *concept structuring technique* called *typology analysis* (see Chap. 2). Essentially, a typology inter-relates simple terms, ideas and concepts in order to create and explore the more complex concepts which are *compounded* out of these simple concepts. A typology classifies the different *types* of something, according to a set of common characteristics or attributes. This is why typologies or typological models are sometimes called "attribute fields".

However, typologies usually only work with two dimensions – i.e. they relate the attributes of two issues: for instance, Jung's scheme of pitting two personality attributes (introvert-extrovert *vs.* rational-irrational) against each other, giving four possible *personality types*. Even typologies involving three dimensions start to become unwieldy.

What we needed to develop was an expanded form of typology analysis which could treat any number of dimensions. With this in mind, I started to explore different methods of representing multi-dimensional typologies in ways that would make sense – conceptually and visually – and could be exploited by the computer.

When I began working on this in the early 1990s, I actually thought that I might be doing something new. If professional typologists (traditionally sociologists) couldn't find a smart way to exploit computers to represent multi-dimensional typologies, then maybe a mathematically oriented anthropologist could. It didn't take long to "get real". Someone had already done it. But when the answer came, it came from a completely different direction than sociology and anthropology.

Extended typology analysis was invented as early as the 1940s by Fritz Zwicky, professor of astrophysics at the California Institute of Technology – the famous Caltech in Pasadena. These days, most people have never heard of Zwicky, but 50 years ago he was a relatively well-known scientific personality in astronomy. He developed the first galaxy catalogue, coined the term *supernova* and was the first to hypothesize the existence of *neutron stars*. He is also regarded by some as being the father of the modern jet engine (see Chap. 9 for a short biography of Zwicky).

Zwicky developed a general form of non-quantified, dimensional analysis in order (*inter alia*) to categorize and hypothesize new types of astrophysical objects, to develop jet and rocket propulsion systems, and to study the legal aspects of space travel. He called this *morphological research*. Later on, it became *morphological analysis*. However, since there are a number of other scientific disciplines that use this term for specific areas of study (e.g. botany, geology, linguistics,), I started calling it *general* morphological analysis (GMA).

Zwicky, and subsequent practitioners of GMA, did it "by hand", or with only rudimentary computer support. This places severe restrictions on the number and range of the dimensions that can be employed (5–6 dimensions is already pushing it). It is also time consuming and prone to errors. But most important of all, without adequate, dedicated computer support, one cannot properly do morphological modelling in a real-time workshop setting with subject matter specialists (SMS). This is crucial: it is the collective creativity which comes out of facilitated group workshops which is at the heart of developing really useful, innovative morphological models concerning wicked problems.

In 1995, I began to develop software support for GMA with this goal in mind: to be able to develop non-quantified, interactive ("what-if") inference models in a real-time workshop setting with subject-matter-specialist and stakeholder groups.

However, there is a lovely Catch 22 situation involved. On the one hand, you will not really understand what types of functionalities and flexibilities you will need in your workshop oriented GMA software until you start running GMA workshops. But, on the other hand, you will find it really difficult to run proper GMA workshops without dedicated (well-thought-out) GMA software.

This is a classical *bootstrapping* problem, and anyone who has started his or her own business from scratch knows exactly what this is about.

Let's say that you work for a commercial organisation that, naturally, has to make a profit and wants satisfied customers; or, alternatively, you work in a bureaucratic organisation that practices extreme risk avoidance (which most bureaucratic organisations do). In either case, if you go out and "burn" five clients in a row in trying to introduce, understand and develop a new method or product – then you are either going to stop doing this on your own accord, or someone else is going to pull the plug on you.

Nobody pulled my plug. I turned out to be the proverbial "right person in the right place at the right time". Above all, I had the "right" colonel supporting the whole effort (it always seems to be a *colonel!*). For 2 years I was allowed to engage dozens of defence study groups in the bootstrapping process of learning how to conduct GMA workshops, and learning what software requirements should go along with this. Thus the GMA software was developed *in parallel* and *in interaction* with the development of the facilitation and modelling techniques required for GMA workshops.

It was an authentic "theory and practice" bootstrapping process. There have been several attempts to develop GMA software "back-office", as purely intellectual products. The ones I have tested are strangely inflexible and practically useless in a real-time workshop setting.

So what's the bottom line here? What is it that *General Morphological Analysis* is actually good for?

GMA is good for the process of stakeholders learning to understand the complex issues and interrelations of the *wicked problems* they are confronted with, and for helping these stakeholders to better understand each other's positions and rationales concerning these issues. The single take-home message is this: never begin any project involving complex policy and planning issues, and complex stakeholder positions, without first engaging the stakeholders (and related subject specialists) in *at least* a 2-day morphological analysis workshop. You will be very glad that you did so!

Almost all of the clients that I have worked for during the past 15 years have ended up saying more or less the same thing: "Boy, are we glad we did a morphological analysis of this problem area before we started spending time and money on it. We didn't really have an adequate understanding of the actual problem space".

As far as I know, this is the first book to more or less comprehensively treat computer aided General Morphological Analysis. It has this distinction by default: the discipline is only 15 years old, and I and my colleagues are the only ones to have amassed enough client based projects employing GMA (more than 100) to be able to draw any general conclusions about the method. Still, it is only a first, tentative attempt to put together some sort of systematic picture of the work that has been done. It is my sincere hope that it will help to generate further interest in General Morphology, and result in more and better research papers, and books, on the subject.

We are not many practitioners in the world at this point – maybe a couple of dozen. These are spread out over the globe in e.g. England, France, Holland, Australia, New Zealand, Korea, the U.S.A. and (of course) Sweden. I have personally introduced and trained facilitators in GMA in Holland, England, Singapore, the Republic of South Africa and the U.S.

Concerning the composition of this book, some of the text is based upon earlier articles and presentations of GMA. To avoid accusations of *self-plagiarism* (i.e. reusing one's own old texts without informing the reader), these earlier sources are footnoted for each section.

The reason for using these earlier texts is simple: I have been writing about and presenting GMA for 15 years and I cannot substantially improve upon some of these descriptions and explanations. It would be silly for me to attempt to completely reformulate everything again. Also, I do not think that "self plagiarism" is all that much of an issue. As the composer G. F. Händel said to someone who accused him of using the same musical theme in several different works: "How many good ideas do you think you get in a lifetime, sonny"?

Likewise, some of the case studies presented have been reported in earlier conference papers. These are also noted in the "Case Studies" section.

I want to thank the following people for their help and support:

- Gunilla Ritchey (my wife, and the love of my life).
- My former colleagues at FOI, especially Maria Stenström, who supported and got involved in GMA despite the risk of being ostracized by a conservative bureaucratic leadership and die-hard "quants".
- The *Colonel* (who will remain unnamed) who supported the bootstrapping process of GMA in the middle of the 1990s, and who never lost confidence in its successful application.
- All of my GMA facilitator apprentices whose enthusiasm, and courage, has been an inspiration in itself. (Yes, it takes courage to learn to facilitate GMA workshops.)
- Jan R., my friend and colleague now residing in New Zealand, for his support and promotion of the morphological approach.
- Bruce Garvey and Nasir Hussain – my friends and partners at the Strategy Foresight Partnership in London – for their support and suggestions.
- The curator of the Fritz Zwicky Foundation in Glarus, Switzerland, for allowing me free accesses to the Zwicky Archives there – a most wonderful experience.

Finally, I am honored to have this book published by Springer, not the least because it was Springer that published Fritz Zwicky's first book on Morphological Research some 44 years ago (Zwicky & Wilson, 1967). I think that Fritz would have appreciated this also.

Chapter 2
General Morphological Analysis (GMA)

"... within the final and true world image everything is related to everything, and nothing can be discarded a priori as being unimportant." (Zwicky, 1969)

"Morphological analysis is simply an ordered way of looking at things." (Zwicky, 1948a)

2.1 What's the Problem?

Analyzing and modelling complex social, organisational and political (i.e. policy driven) systems presents us with a number of difficult methodological problems. Firstly, many of the factors involved are not meaningfully quantifiable, since they contain strong social, political and cognitive dimensions. Secondly, the uncertainties inherent in such problem complexes are in principle non-reducible, and often cannot be fully described or delineated. This includes both so-called *agonistic uncertainty* (conscious, self-reflective actions among competing actors) and *non-specified uncertainty* (for instance, uncertainties concerning what types of scientific and technological discoveries will be made in the future).

Finally, the extreme non-linearity of social systems means that literally everything depends on everything else. What might seem to be the most marginal of factors can, under the right historical circumstances, become a dominating force of change. All of this means that traditional quantitative methods, mathematical (functional) modelling and simulation (in the sense of attempting to predict how things are *actually* going to "work out"), are relatively useless.

An alternative to formal (mathematical) methods and causal modelling is a form of non-quantified modelling relying on *judgmental processes* and *internal consistency*, rather than causality. Causal modelling, when applicable, can – and should – be used as an aid to judgement. However, at a certain level of *complexity*[1]

[1]See Chap. 10: complexity (self-referential systems).

This chapter is based on earlier articles published from 1998 and onwards, including Ritchey (1998, 2002, 2006a).

DOI 10.1007/978-3-642-19653-9_2,

(e.g. social, political and cognitive processes), judgement must often be used – and worked with – more or less directly. The question is: How can judgmental processes be put on a sound methodological basis?

Historically, scientific knowledge develops through cycles of analysis and synthesis: every synthesis is built upon the results of a proceeding analysis, and every analysis requires a subsequent synthesis in order to verify and correct its results (Ritchey, 1991). However, analysis and synthesis – as basic scientific methods – say nothing about a problem having to be quantifiable.

Complex social-political problem fields can be analysed into any number of non-quantified variables and ranges of conditions. Similarly, sets of non-quantified conditions can be synthesised into well-defined relationships or configurations, which represent "solution spaces". In this context, there is no fundamental difference between quantified and non-quantified modelling (see Chap. 6).

General Morphological analysis (GMA) is a method for structuring and investigating the total set of relationships contained in multi-dimensional, non-quantifiable, problem complexes. It was originally developed by Fritz Zwicky, the Swiss astrophysicist and aerospace scientist based at the California Institute of Technology (see Chap. 9).

Zwicky applied this method to such diverse fields as the classification of astrophysical objects, the development of jet and rocket propulsion systems, and the legal aspects of space travel and colonization. He founded the Society for Morphological Research and advanced the "morphological approach" for some 40 years, between the early 1930s until his death in 1974.

More recently, morphological analysis has been applied by a number of researchers in the USA and Europe in the fields of policy analysis and futures studies (see References). In 1995, advanced computer support for GMA was developed at the Swedish Defence Research Agency (for a description, see Ritchey, 2003b). This has made it possible to create non-quantified inference models, which significantly extends GMA's functionality and areas of application (see Ritchey, 1997–2009). Since then, more than 100 projects have been carried out using computer aided morphological analysis, for structuring complex policy and planning issues, developing scenario and strategy laboratories, and analyzing organizational and stakeholder structures.

This Chapter will continue with a history of morphological methods, a description of the modelling processes itself, and an example concerning the modelling of an *organisational structure*.

2.2 Short History of Morphological Methods

The term *morphology* comes from classical Greek (*morphê*) and means the study of shape or form. Morphology is concerned with the structure and arrangement of parts of an object, and how these *conform* to create a whole or Gestalt. The "object" in question can be a physical or biological system (e.g. an organism, an anatomy or an ecology), a social system (e.g. an organisation, institution or society) or a mental system (e.g. linguistic forms, concepts or systems of ideas).

Today, morphology is associated with a number of scientific disciplines in which formal structure is a central issue. In biology it is the study of the shape or form of organisms. In linguistics, it is the study of word formation. In geology it is associated with the characteristics, configuration and evolution of rocks and landforms.

The first to use the term *morphology* as an explicitly defined scientific method was J.W. von Goethe (1749–1832). Goethe introduced the term to denote the principles of formation and transformation of organic bodies. Concentrating on form and quality, rather than function and quantity, this approach produced generalizations about the combinatorial logic of biological structures. Of central importance was the idea of the *morphotype*; that is, a structural or organisational principle which can be identified and studied through comparative anatomy.

This early theoretical morphology was eventually eclipsed by Darwinian evolutionary theory in the late nineteenth century. With the exception of the works of William Bateson (1896) and D'Arcy Thompson (1917), it remained obscure until the Modern Synthesis in evolutionary biology began to treat Darwinian evolution from at the level of genes, phenotypes and populations. The present literature in theoretical morphology is now quite extensive (McGhee, 1999).

It is important to note, that Goethe developed morphology with the expressed purpose of methodologically distancing the life sciences from the then reining paradigm in *Naturwissenschaft*, i.e. classical (Newtonian) mechanics. However, this methodological shift was exactly what was needed in another area, which was even less disposed to such a paradigm: the emerging disciplines of sociology and psychology. Theoretical morphology was thus carried over into the *Geisteswissenschaft* of Classical German Sociology – represented by Wilhelm Dilthey (1833–1891) (Dilthey, 1989) and Max Weber (1864–1920) (Weber, 1949). More specifically, morphology and morphotypes became typology and ideal types.

A typology (the Greek word *typos* originally meant a hollow mould or matrix) is a very simple morphological model based on the possible combinations obtained between a few (often two) variables, each containing a range of discrete values or states. Each of the possible combinations of variable-values in the typological field is called a *constructed type*. Typologies abound, especially in the sociological literature, and typology analysis is virtually a discipline in itself (Bailey, 1994; Doty & Glick, 1994). The simplest and most common form of a typology is the ubiquitous *four-fold table* (or a so-called MBA 2×2), which pits two variables against each other, each variable containing two values or states.

The type-concept was not created by Weber. It was already well established methodologically by Goethe in his conception of morphotypes. However, by employing typologies as a method for formulating sociological and social philosophical categories, Weber simplified, generalised and popularised typology analysis as a simple *concept-structuring method* applicable to virtually any area of investigation.

Although typological fields are certainly not restricted to two dimensions or simple binary relations, there are severe limits to the complexity of the classical typological format. Visually, a typology utilizes the dimensions of physical space to represent its variables. Each of the constructed types lies at the intersection of two or more coordinates. However, the number of coordinates that can be represented in physical space ends at *three*. Typologies of greater dimensions – representing hyperspaces – usually get around this problem by embedding variables within each other. However, such formats quickly become difficult to interpret, if not hopelessly unintelligible. There are, however, other ways to represent hyperspaces.

In the late 1940s, Fritz Zwicky, the Swiss astrophysicist and aerospace scientist based at the California Institute of Technology (Caltech), proposed a generalized form of morphological analysis:

> "Attention has been called to the fact that the term morphology has long been used in many fields of science to designate research on structural interrelations - for instance in anatomy, geology, botany and biology. . . . I have proposed to generalize and systematize the concept of morphological research and include not only the study of the shapes of geometrical, geological, biological, and generally material structures, but also to study the more abstract structural interrelations among phenomena, concepts, and ideas, whatever their character might be." (Zwicky, 1969, p 34)

In general morphology, the problem of representing – and visualising – more than three dimensions is overcome by placing the variables in columns beside each other, their value ranges listed below them. This is called a *morphological field.* A particular constructed morphotype (called a *field configuration*) is designated by selecting a single value from each variable (see Fig. 2.1).

Zwicky published a number of articles applying morphology to the classification of astrophysical objects (Zwicky, 1948a) and to the development of jet and rocket

Character of Chemical Reactions	Method of Thrust Augmentation 1	Method of Thrust Augmentation 2	Physical State of Propellants	Operating Mode of Propulsive Power Plant	Reactivity or Reaction Speed of the Propellants
Self-contained - carries all chemicals necessary for activation and operation.	No motion	No thrust augmentation	Gaseous state	Continuous operation	Propellants are self-igniting
If air-propelled, carries only fuel and uses atmospheric oxygen.	Translatory motion	Internal thrust augmentation	Liquid state	Intermittent (pulsating) operation	Artificial ignition is necessary
If propelled through or over water, uses water as propellant reacting with an on-board water-reactive chemical.	Rotary motion	External thrust augmentation	Solid state		
If propelled through or over the earth may use earth as propellant reacting with an on-board earth-reactive chemical.	Oscillatory motion				

Fig. 2.1 Zwicky's "propulsive system morphology" from 1947, containing six dimensions (parameters) and 576 (4 × 4 × 3 × 3 × 2 × 2) formal configurations – one displayed

propulsion systems (Zwicky, 1947). He also published a more general article on the "morphological method of analysis and construction" (Zwicky, 1948b) and later wrote a book on the subject (Zwicky, 1969). His morphological astronomy lead to a number of hypotheses and later discoveries, but remained more or less specific to astrophysics. His work on jet propulsion systems, however, had a wider impact in the area of engineering design.

In 1962, in a paper presented at a conference on engineering design methods in London, Norris (1963) proposed that the morphological approach should be turned into a full-fledged engineering design method utilising computers (!), in order to systematically separate and collate different design solutions. Some authors saw even wider applications. Ayres (1969) pointed out how morphological analysis could be employed to systematically generate scenarios. He cited the work on future, non-national nuclear threats by Theodore Taylor (1967) at the Stanford Research Institute.

In 1975, Müller-Merbach (1976) of the University of Darmstadt wrote an article for *Operational Research* titled "The Use of Morphological Techniques for OR-Approaches to Problems". There he pointed out that general morphology is especially suitable for operational research, not the least because of the growing need for operational analysts to be part of the *problem formulation process*, and not simply a "receiver" of pre-defined problems.

In a more specific context, Rhyne (1971, 1981) – also associated with the Stanford Research Institute – picked up on Taylor's earlier work and began to apply a somewhat restricted form of morphological analysis as a scenario development technique. (In order to generate new interest in the method, Rhyne packaged it under the esoteric name of "field anomaly relaxation" (FAR), a term borrowed from mechanical engineering [personal communication].) During the 1990s he continued to write about its potential as a systematic approach to futures projections (Rhyne, 1995a, 1995b).

Finally, in the early 1990s, Geoff Coyle, then working at the Royal Military Collage of Science in Swindon, discovered Rhyne's work and promoted morphological analysis as one of a number of structured techniques for scenario development (Coyle et al., 1995, 1996).

Unfortunately, GMA has been written about and discussed far more that it has actually been used in "real" client-based projects. One of the principle reasons for this, I believe, is that it has been mostly carried out by hand or with only rudimentary computer support. Employing GMA in this way is not only extremely difficult, time consuming and prone to errors; it severely limits the number and range of parameters that can be employed. Since the number of configurations (i.e. formal solutions) in a morphological field increases exponentially (or in a *factorial* manner) with the number of parameters applied to it, working with as few as six or seven variables becomes a considerable task. Thus, until recently, GMA has usually been carried out as a relatively simple form of attribute listing with internal consistency checks.

In 1995, my colleagues and I at the Department for Technology Foresight and Assessment at Totalförsvarets Forskningsinstitut (FOI – the Swedish Defence Research Agency in Stockholm) realized that general morphological analysis would never reach its full potential without *dedicated, highly flexible, workshop oriented* computer support. The system we began developing then, and which is presently in its fifth programming generation, fully supports the analysis-synthesis cycles inherent in GMA and makes it possible to create morphological ("what-if") inference models. During the past 15 years, GMA has been utilised in more than 100 client-based projects, for structuring complex policy and planning issues, developing scenario and strategy laboratories, and analysing organisational and stakeholder structures.

2.3 Morphological Modelling

Since we will frequently be using the terms *model* and *modelling*, it is best to get these concepts defined at the outset. Although there is no concise, unanimously agreed upon *general definition* of a (scientific) model, for the purpose of this study, we posit the following three conditions as necessary and sufficient for a *minimal definition* of a (scientific) model[2]:

- A model must contain two or more constructs that can serve as variables, i.e. dimensions which can support a range of states or values. [In morphological modelling we call these the model's *parameters*. We define a *parameter* as being one of a set of measurable factors that define a system and determine its behaviour, and which can be varied in an experiment – including a *Gedanken experiment*.]
- One must be able to establish relationships (causal, statistical, logical, etc.) between the states or values of the different parameters. [In morphological modelling the relationships are predominantly "logical" in the sense that they concern consistency, coherence or co-existence.]
- Inputs can be given, and outputs obtained. [In morphological modelling, this is achieved by (temporarily) designating one or more parameters as independent variables (inputs) and realising the results on the remaining variables (outputs)].

Morphological modelling is simply a non-quantified, discrete-variable application of these requirements. As discussed above, the process goes through cycles of analysis and synthesis – the basic procedure for developing all (scientific) models (Ritchey, 1991).

The analysis phase begins by identifying and defining the most important dimensions of the problem complex to be investigated. Each of these dimensions is then given a range of relevant values or conditions. Together, these make up the variables or parameters of the problem to be structured. A morphological field is

[2]Some might object that this definition excludes the classical "influence diagram" as a *scientific* model. But I think it significant that influence diagrams are called *diagrams*, and not *models*. These diagrams represent nodes as black boxes with "arrows" of influences depicted between the nodes. They allow for no variability or inference. However, what is, and what is not, to be considered a *scientific model* is a matter of convention, as long as our definitions are clear and we apply them consistently.

constructed by setting the parameters against each other, in parallel columns, representing an n-dimensional configuration space. A particular constructed "field configuration" (morphotype) is designated by selecting a single value from each of the variables. This marks out a particular state or (formal) solution within the problem complex (as in Fig. 2.1).

Ideally, one would examine all of the configurations in the field, in order to establish which of them are possible, viable, practical, interesting and so forth, and which are not. In doing so, we mark out in the field a relevant "solution space". The solution space of a Zwickian morphological field consists of the subset of configurations which satisfy some set of criteria – one of which is *internal consistency.*

However, typical morphological fields of six to ten variables can contain between 50,000 and 5,000,000 formal configurations, far too many to inspect by hand. Thus, the next step in the analysis-synthesis process is to examine the internal relationships between the field parameters and *reduce* the field by identifying, and weeding out, all mutually contradictory conditions.

This is achieved by a process of *cross-consistency assessment* (CCA). All of the parameter values in the morphological field are compared with one another, pair-wise, in the manner of a cross-impact matrix (Fig. 2.2). As each pair of conditions is examined, a judgment is made as to whether – or to what extent – the pair can coexist, i.e represent a consistent relationship. Note that there is no reference here to direction or causality, but only to mutual consistency. Using this technique, a typical morphological field can be reduced by up to 90 or even 99%, depending on the problem structure. (Certain types of *scenario fields* are an exception, as will be discussed below.)

		Character of				Method of Th				Method of			State of P			Opera	
		Self-contained	If air-propelled	Water propelled	Earth propelled	No motion	Translatory motion	Rotary motion	Oscillatory motion	No augmentation	Internal augmentation	External augmentation	Gaseous state	Liquid state	Solid state	Continuous	Intermittent
Method of Thrust 1	No motion																
	Translatory motion																
	Rotary motion																
	Oscillatory motion																
Method of Thrust 2	No augmentation																
	Internal augmentation																
	External augmentation																
State of Propellants	Gaseous state																
	Liquid state																
	Solid state																
Operating Mode	Continuous																
	Intermittent																
Reactivity	Self-igniting																
	Artificial ignition																

Fig. 2.2 The cross-consistency matrix (CCM) for the field in Fig. 2.1

There are two principal types of inconsistencies involved here: purely *logical contradictions* (i.e. those based on the nature of the concepts involved); and *empirical constraints* (i.e. relationships judged be highly improbable or implausible on empirical grounds). *Normative constraints* can also be applied, although these must be used with great care, and clearly designated as such. In general, we first want to distinguish between what is *possible* and *not possible* (or not plausible), before going on to consider normative issues. (Although, as we shall see, some models are *predominately normative* in character.)

This technique of using pair-wise consistency relationships between conditions, in order to weed out internally inconsistent configurations, is made possible by the principle of dimensionally inherent in the morphological approach. While the number of configurations in a morphological field grows exponentially with each new parameter, the number of *pair-wise relationships between conditions* grows only as a quadratic polynomial – more specifically, in proportion to the triangular number series (see Chap. 6). Naturally, there are practical limits reached even with quadratic growth. However, a morphological field involving as many as 100,000 formal configurations can require no more than few hundred pair-wise consistency assessments in order to create a solution space.

When this solution space is synthesized, the resultant morphological field becomes an interactive inference model, in which any parameter (or multiple parameters) can be selected as "input", and any others as "output". Thus, with proper computer support, the field becomes a *conceptual laboratory* for exploring knowledge bases and solution requirements, testing assumptions and interventions, and spotting potential *unintended consequences* – which are one of the main outcomes of intervening into *wicked problems*.

GMA employs *facilitated group interaction* as a central feature of the modelling process, since we are not only structuring a complex problem, but creating among the participants shared concepts and a common modelling framework. What is essentially a process of collective creativity is best facilitated in dialogue between participants, rather than each participant addressing an "assembly". For this reason, we have found it best to work with subject specialist groups of no more than six to seven persons. If a wider knowledge base is required, one can either bring specialized competence into specific group sessions, or work in parallel groups (see Chap. 7).

Depending on the level of ambition (e.g. how many different models a client wishes to develop; the complexity of the models; and the number of groups involved) a modelling job can take between two and ten workshops days. We utilize two facilitators per workshop group. These alternate between, on the one hand, facilitating the group process as such and, on the other hand, tending the computer, recording and reflecting. Virtually all of the work is done in the workshop setting, with little back-office or software preparation time required.

Also, the software is designed to facilitate project documentation during the workshop sessions themselves. The models that are generated during these sessions belong to the client, who is provided with software and documentation to run and maintain them.

2.4 Example: Organisational Structure

A simple example of a morphological model may suffice to illustrate the principles of the method. It is drawn from a project done in the late 1990s for the Swedish National Defence Research Agency (FOI) concerning future *Organisational structure*. In fact three models were developed for the project: *Organisational structure*, *Markets and clients* (see Case Studies) and *Security and legal issues*. (*Note*: the model shown here is a truncated version of the original model. It is employed here only as a pedagogical example.)

With the end of the Cold War, Swedish defence research (as in many other countries) began to develop into broader areas of interest than simply territorial or invasion defence. Also, with changing threat perceptions, there were clear budgetary issues afoot (i.e. budgets were going to be cut!). How could a predominately national defence oriented organisation like FOI reform or re-invent itself to cope with new post-Cold War developments.

The first problem is to identify and properly define the dimensions of the problem – that is to say, the relevant *issues* or parameters involved. These included organisational and leadership types, client sectors, products and employee profiles – all at a relevant level of abstraction. One of the advantages of GMA is that there are no formal constraints to mixing and comparing such different types of issues. On the contrary, if we are really to get to the bottom of an organisational or policy problem, we must treat all relevant issues *together*.

Secondly, for each issue (parameter), a spectrum of "values" must be defined. These values represent the possible, relevant states or conditions that each issue can assume, for the particular study at hand.

The process embodied in these two "steps" is an iterative one, much like the iterative ups-and-downs illustrated in Conklin's diagram describing the time line of structuring "wicked problems" (see Chap. 3).

The morphological field for the organisational structure model is shown in Fig. 2.3. It contains 186, 624 possible configurations – which is simply the product of the number of values under each parameter.

The next (iterative) step in the analysis-synthesis process is to reduce the total set of (formally) possible configurations in the morphological field to a smaller set of internally consistent configurations representing a "solution space". (This is what Zwicky called the *principle of contradiction and reduction*, and what we call a "Cross-Consistency Assessment" (CCA).)

This reduction allows us to concentrate on a manageable number of internally consistent configurations. With dedicated software, we can designate inputs, define drivers and examined resultant output configurations as elements of scenarios or specific strategies in a complex policy space. Figure 2.5 shows the model designated with three inputs (grey).

Organisation TYPE	Leadership culture	Dominant buyer structure	Dominate product/ service	Co-operation strategies	Principle Employee profile	Main employee incentive
Official state agency	Bureaucratic hierarchy	Ministry dominated	Process + method support	Outside help when needed	Life-long service	Money
Government owned enterprise	Strong scientific leadership	Military and material dominated	Soft studies	Joint ventures	Career researcher	Managerial career
Academy	Marketing division leadership	Defence Industry	Hard studies	Consultant purchasing	Development engineer	Pleasure in one's work
Trade institute	Umbrella management	Civilian agencies	Basic research	Mediator only	"Consultant"	Educational motivation
Consultant firm	"Gatekeeper"	Private markets (national)	Testing, construction		Entrepreneur	Titles, specialist career
"Learning organisation"	Skunk-works	International markets	Second opinion		Elite troops	Organisation gives status

Fig. 2.3 One of the organisational development models produced for the Swedish Defence Research Agency in the late 1990s

		Organisation						Leadership						Dominant						Dominate						Co-operatio				Principle					
		Official state agency	Governmentowned	Academy	Trade institute	Consultant firm	earning organisation"	ureaucratic hierarchy	Strong scientific	Marketing division	mbrella management	"Gatekeeper"	Skunk-works	Ministry dominated	Military and material	Defence Industry	Civilian agencies	Private markets	Internationalmarkets	Process + method	Soft studies	Hard studies	Basic research	Testing, construction	Second opinion	Outside help	Joint ventures	onsultant purchasing	Mediator only	Life-long	Career researcher	evelopment engineer	"Consultant"	Entrepreneur	Elite troops
Leadership culture	Bureaucratic	F	-	X	F	X	X																												
	Strong	X	X	F	X	X	-																												
	Marketing	K	F	X	F	F	-																												
	Umbrella	X	-	-	X	F	F																												
	"Gatekeeper"	F	K	-	K	X	X																												
	Skunk-works	X	X	-	X	-	F																												
Dominant buyer structure	Ministry	F	K	-	F	K	K	F	X	K	-	-	-																						
	Military and	F	-	-	F	-	K	F	X	K	K	-	-																						
	Defence	K	-	-	-	-	-	K	-	F	-	-	-																						
	Civilian	-	-	K	K	-	-	-	K	-	F	-	-																						
	Private markets	X	-	X	X	-	F	X	-	F	F	-	-																						
	International	X	-	-	X	-	F	X	-	F	F	-	-																						
Dominate product/ service	Process + method	F	-	K	F	F	F	-	-	-	-	-	K	-	F	X	-	-	-																
	Soft studies	F	-	F	F	K	F	-	-	-	-	-	-	F	-	K	F	-	-																
	Hard studies	-	-	F	F	F	F	-	-	-	-	-	-	-	-	F	-	-	-																
	Basic research	K	X	F	K	X	F	K	F	K	-	F	-	X	X	X	X	X	-																
	Testing, construction	-	-	-	K	-	F	-	K	-	-	-	-	X	-	-	-	-	-																
	Second opinion	F	K	K	F	-	F	-	-	K	-	-	K	F	X	X	-	X	-																
Co-operation strategies	Outside help	-	-	-	-	-	F	-	-	-	-	-	-	-	-	-	-	-	-	-	-	-	-	-	-										
	Joint	-	-	-	-	K	F	K	-	-	K	-	X	K	K	-	-	-	-	K	-	-	-	-	-										
	Consultant	-	-	X	X	K	X	-	K	-	K	-	-	K	K	-	-	-	-	X	-	-	-	K	X										
	Mediator only	-	X	X	X	-	-	K	X	-	X	K	X	X	K	-	-	-	-	X	-	-	X	X	X										
Principle Employee profile	Life-long	F	-	F	F	X	X	F	-	X	X	F	X	-	-	K	K	K	X	-	-	-	-	-	-	-	-	-	-						
	Career researcher	-	K	F	-	K	-	-	F	X	-	F	F	-	-	-	-	-	-	-	-	-	F	-	-	-	-	-	X						
	Development engineer	K	-	-	K	-	-	-	-	F	-	-	F	K	-	-	-	-	-	K	X	-	X	-	X	-	-	-	X						
	"Consultant"	F	-	K	X	F	K	-	K	F	F	K	X	-	-	-	-	-	-	-	-	-	X	-	-	-	-	-	-						
	Entrepreneur	-	-	F	K	-	F	X	X	F	F	X	F	-	-	F	F	F	F	-	-	-	-	-	-	-	-	-	-						
	Elite troops	K	K	F	K	-	F	X	F	-	-	-	F	-	-	-	-	F	F	-	-	-	-	-	-	-	-	-	-						
Main employee incentive	Money	K	F	X	K	F	-	K	-	F	-	-	-	-	-	-	-	-	-	-	-	-	-	-	-	-	-	-	-	K	K	F	F	-	F
	Managerial career	F	-	F	-	K	K	F	K	-	-	-	K	-	-	-	-	-	-	-	-	-	-	-	-	-	-	-	F	-	-	K	K	-	-
	Pleasure	-	-	F	-	-	F	-	-	-	F	-	F	-	-	-	-	-	-	-	-	-	F		-	-	-	-	-	-	F	F	-	F	F
	Educational motivation	-	-	-	F	F	-	K	-	-	-	-	-	-	-	-	-	-	-	-	-	-	-	-	-	-	-	-	-	X	-	-	-	-	-
	Titles, specialist	X	K	F	K	-	-	K	-	-	K	-	K	-	-	-	-	-	-	-	-	-	-	-	-	-	-	-	K	X	F	-	F	-	-
	Organisation status	F	-	F	X	K	F	-	-	-	F	-	F	-	-	-	-	-	-	-	-	-	F	-	-	-	-	-	-	-	F	-	-	F	F

Fig. 2.4 Cross-consistency matrix (CCM) for the organisational development model in Fig. 2.3

Organisation TYPE	Leadership culture	Dominant buyer structure	Dominate product/ service	Co-operation strategies	Principle Employee profile	Main employee incentive
Official state agency	Bureaucratic hierarchy	Ministry dominated	Process + method support	Outside help when needed	Life-long service	Money
Government owned enterprise	Strong scientific leadership	Military and material dominated	Soft studies	Joint ventures	Career researcher	Managerial career
Academy	Marketing division leadership	Defence Industry	Hard studies	Consultant purchasing	Development engineer	Pleasure in one's work
Trade institute	Umbrella management	Civilian agencies	Basic research	Mediator only	"Consultant"	Educational motivation
Consultant firm	"Gatekeeper"	Private markets (national)	Testing, construction		Entrepreneur	Titles, specialist career
"Learning organisation"	Skunk-works	International markets	Second opinion		Elite troops	Organisation gives status

Fig. 2.5 Organisational structure model with three inputs (*grey*) and resultant output cluster (*black*)

2.5 Conclusions

General Morphological Analysis, including the process of "Cross-Consistency Assessment" (CCA), is based on the fundamental scientific method of alternating between analysis and synthesis. For this reason, it can be trusted as a useful, non-quantified method for investigating problem complexes, which cannot be meaningfully treated by formal mathematical methods, causal modelling and simulation.

However, as a non-quantified modelling method, GMA has several advantages over less structured approaches. Zwicky called it "totality research" which, in an "unbiased way attempts to derive all the solutions of any given problem". It may help us to discover new relationships or configurations which might be overlooked by other – less structured – methods. Importantly, it encourages the identification and investigation of *boundary conditions*, i.e. the limits and extremes of different contexts and factors.

It also has definite advantages for scientific communication and – notably – for group work. As a process, the method demands that parameters, conditions and the issues underlying these be clearly defined. Poorly defined parameters become immediately (and embarrassingly) evident when they are cross-referenced and assessed for internal consistency. This provides for a good deal of in-built "garbage detection", since these assessments simply cannot be made until the morphological field is well defined and the working group is in agreement about these definitions.

This type of *garbage detection* is something that policy analyses and futures studies certainly need more of.

Also, both the *formulation* of the morphological field itself, and the assessments put into the cross-consistency matrix, represent a fairly clear "audit trail", which makes the judgmental processes inherent in GMA relatively traceable, and – in a certain sense – even reproducible. We have run trials in which identical morphological fields were presented to different groups for cross-consistency assessment. Comparing the results, and bringing the groups together to discuss diverging assessments, helps us to better understand the nature of the policy issues involved, and also tells us something about the effects of group composition on the assessments.

One final note on morphological modelling: GMA is a fundamental method and very general procedure. It leaves open a number of questions about dependencies, independent variables, what is "input" vs. "output", what different types of consistencies are employed, etc. To attempt to impose one or another of these issues *beforehand* on the modelling process, or in the software applications, would be a huge mistake. It is the very *open nature* of these questions that allows for the creative exploration of the modelling space. Not "fixing" these issues beforehand gives us the possibility of free but disciplined creativity – what Bernhard Reimann called the *poetry of hypothesis*.

Before going on to a more detailed description of the GMA procedure (Chap. 4), we can first take a look at the main type of problem complex for which GMA was initially developed – i.e. *wicked problems* and *social messes*.

Chapter 3
Wicked Problems and Genuine Uncertainty

3.1 Background

In 1973, Horst Rittel and Melvin Webber, both urban planners at the University of Berkley in California, wrote an article for *Policy Sciences* with an astounding title: *Dilemmas in a General Theory of Planning*. In this article, the authors observed that there is a whole realm of social planning problems that cannot be successfully treated with traditional linear, analytical approaches. They called these *wicked problems*, in contrast to *tame problems*.

Just a year later, in his book "Redesigning the Future", Ackoff (1974) essentially put forward the same concept, although in less detail, which he called *messes* or *unstructured reality* – and which later became known as "social messes" (Horn 2001).

The best way to appreciate the nature of *wicked problems* is to compare them to *tame problems* (comp. Conklin, 2001). A tame problem

- Has a relatively well-defined and *stable problem statement.*
- Has a *definite stopping point*, i.e. we know when the solution or a solution is reached.
- Has a solution which can be *objectively evaluated* as being right or wrong.
- Belongs to a *class of similar problems* which can be solved in a similar manner.
- Has solutions which can be *tried and abandoned.*

The definitive model for a tame problem is a relatively well-defined *engineering problem*: building a bridge, designing a circuit, putting a person on the moon. Such problems are not necessarily "simple" – they can be very complicated indeed. However, their very property of being well defined and stable inherently defines a

---

Some of the text in this section is taken from two earlier sources, including an address to the Royal Institute of Technology in Stockholm, and a paper presented to SRA (Society for Risk Analysis) Conference in Paris, 2004: "Modelling Society's Capacity to Manage Extraordinary Events".

T. Ritchey, *Wicked Problems – Social Messes*, Risk, Governance and Society 17,
DOI 10.1007/978-3-642-19653-9_3,

range of solution concepts. Also, their stability (at least in the relatively short-term) means that they are essentially the same problem today as they were yesterday, and will be more or less the same problem tomorrow.

Wicked problems (WPs) are completely different. WPs are ill-defined, ambiguous and associated with strong moral, political and professional issues. They are *subjective* and strongly stakeholder dependent: there is often little consensus about what the problem actually *is*, let alone how to resolve it. Above all, WPs won't keep still: they are sets of complex, interacting issues evolving in a dynamic social context. Often, new forms of wicked problems emerge *as a result* of trying to understand and solve one of them.

The most obvious wicked problems are complex, long-term social and organisational planning and policy problems. Examples:

- How should we fight the "War on Terrorism?"
- How do we get democracies to emerge from authoritarian regimes?
- What is a good national immigration policy?
- How should scientific and technological development be governed?
- What should we do to deal with crime and violence in our schools?
- How should our organisation develop in the face of an increasingly uncertain future?

> "The classical systems approach ... is based on the assumption that a planning project can be organized into distinct phases: 'understand the problems', 'gather information,' 'synthesize information and wait for the creative leap,' 'work out solutions' and the like. For wicked problems, however, this type of scheme does not work. One cannot understand the problem without knowing about its context; one cannot meaningfully search for information without the orientation of a solution concept; *one cannot first understand, then solve*". (Rittel & Webber, 1973, p. 161) (Emphasis added.)

Conklin (2001) represents the difference in tackling *tame* and *wicked* problems with the following diagrams (Figs. 3.1 and 3.2).

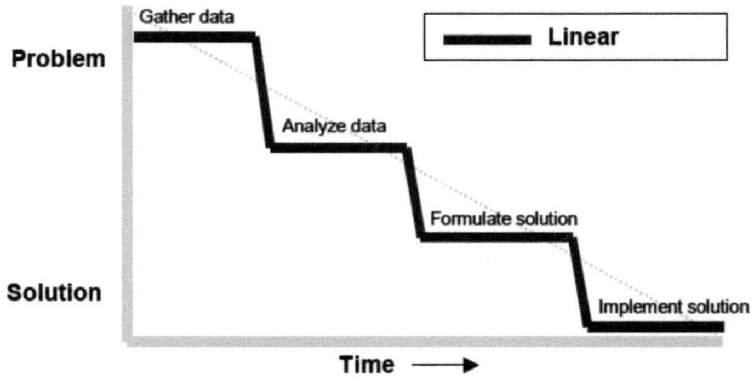

Fig. 3.1 Traditional wisdom for solving complex problems: the "waterfall" (Conklin, 2001)

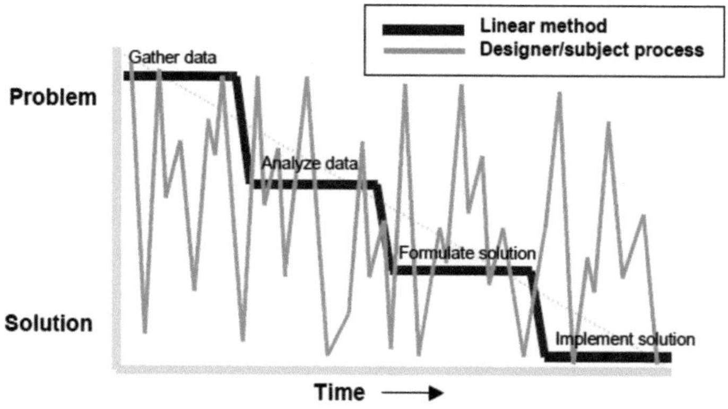

Fig. 3.2 The iterative pattern of working with, and understanding, a wicked problem (Designer/subject process – *grey zig–zags*) (Conklin, 2001)

Conklin continues:

> "... there are two common organizational coping mechanisms that are routinely applied to wicked problems: *studying* the problem, and *taming* it.
>
> While *studying* a novel and complex problem is natural and important, it is an approach that will run out of gas quickly if the problem is wicked. Pure study amounts to procrastination, because little can be learned about a wicked problem by objective data gathering and analysis. Wicked problems demand an opportunity-driven approach; they require making decisions, doing experiments, launching pilot pro-grams, testing prototypes, and so on. Study alone leads to more study, and results in the condition known as 'analysis paralysis,' a Catch 22 in which we can't take action until we have more information, but we can't get more information until someone takes action. ...
>
> ... attempting to tame a wicked problem, while appealing in the short run, fails in the long run. The wicked problem simply reasserts itself, perhaps in a different guise, as if nothing had been done. Or, worse, sometimes the tame solution exacerbates the problem". (Conklin, 2001, p. 10f.)

Of course, problems are *wicked* and *tame* only *a potiori*. In practice there is a sort of gliding scale between *tameness* and *wickedness*. However, there is a set of pretty clear criteria for judging the *degree of wickedness* (so to speak) associated with complex social and organisational planning problems. We will look at Rittel and Webber's ten criteria shortly.

Dilemmas in a General Theory of Planning has been described as a landmark article. I do not disagree with this. When I first discovered it in the early 1990s I was totally fascinated by the issues which it touched upon and by the provocative way in which the authors presented their ideas. Clearly, the authors were challenging us to test our preconceived (or, more correctly, our academically crammed) ideas about the nature of social planning and what we actually mean by *problems* and *solutions*.

However, the "problem" that the concept of *wicked problems* addresses did not suddenly emerge in the late 1960s and early 1970s. WPs are about people, stakeholders, vested interests and politics. As such, they are as old as human society itself. So, if WPs are everywhere and all-the-time, then why is Rittel &Webber's

article referred to as "landmark", and why did at least two separate U.S. based research groups (R&W on the one hand, and Russel Ackoff on the other) start fussing about all this at about the same time.[1]

During a presentation of GMA in the late 1990s, I naively put this question to a group of U.S. security specialists. They practically fell off their chairs with laughter! What was going on? The Vietnam War was going on, along with the "War on Poverty" and the "War on Drugs". All of these "wars" were essentially managed (badly) like huge engineering projects, and all ultimately bogged down or went seriously wrong. Also "going on", as a *reaction* to government policy (or lack thereof), were the Civil Rights Movement, the anti-war demonstrations and the general baby boomer "revolution".

It is no wonder that academics representing social planners and policy professionals sought a new awareness and new modes of explanation. As Rittel & Webber put it in 1973 (I paraphrase): The "publics" are not going to take it any longer, and planners are going to be held accountable for what they do, or don't do. (Wouldn't that have been nice?)

So what *is* the problem that the term "wicked problem" addresses?

The common sense approach to WPs is fairly straight forward: As stated above, WPs are about people and politics; they are *subjective problems*. Everything that has to do with people and society is ultimately *subjective*. People think, reflect, get mad and react. And whatever some philosophers, psychologists or neurobiologists might say about the notion of "free will", most people think that they have it, and, on occasion, act accordingly.

Above all, WPs are about people as *stakeholders*. Stakeholders – competing and cooperating, vying for position, and willing to re-think and change their positions when it suits them – are the epitome of wicked problems. This is why such problems do not have *stable problem formulations*; do not have pre-defined *solution concepts*; and why their course of development *cannot be predicted*.

If you think that all of this is pretty straight forward or self-evident, you are right, and you can skip the rest of this chapter. You don't need it.

Going beyond this "common sense" approach, there is any number of epistemological-ontological issues lurking here. These concern, for instance, the problem of *freedom vs. necessity,* the distinction between *causality* and *determinism* and the relationship between *being and becoming*. However, this is not a philosophical tract and I do not have the intellectual wherewithal to go down this path again. (I wrote my doctoral thesis on this subject 30 years ago and still get a migraine when I think about it.)

---

[1]In fact, Zwicky was already fully aware of the "wicked problem" issue in the 1950s – long before Rittel and Ackoff began writing about it. He stressed that many of the problems suitable to be tackled by GMA require

> "...an integrated view which relates [technical knowledge] to political, psychological and ethical factors... All of these factors add up to a complex task which is beyond the power of ordinary scientific, technical and managerial experts". (Zwicky, 1960, p. 22)

Instead, I want to limit the discussion to two practical *methodological* issues. These are:

- The *etic* and the *emic* approaches to social/cultural research.
- The issue of *genuine uncertainty*.

3.2 The Etic vs. the Emic

The two main pillars of social/cultural research are the so-called *etic* and *emic* approaches. These terms were coined by the linguist Kenneth Pike in the early 1950s (Pike, 1954).[2] However, the concepts they represent were hardly new at that time. The same essential concepts were expressed by Philo of Alexandria in 50 BCE (Wright, 2009), by G. W. Leibniz' (1646–1715) in the relationship between *efficient* and *final* causes (Ritchey, 1983), and by Marx' (often misappropriated) *dialectical method* – i.e. the continual developmental relationship between the ("objective") material world and ("subjective") human apperception of the world.

In studying social systems, the *etic* perspective relies on *extrinsic concepts and categories* that have meaning for scientific observers (e.g. per capita energy consumption, the spread of an epidemic or crime statistics). Professionals ("scientists") studying these phenomena are the primary judges of the validity of an *etic* account. For simplicity sake, call it the *objective* account, in the sense that something is being studied from the outside, as an *object*.

The *emic* perspective focuses on *intrinsic cultural distinctions, perceptions and motivations* that are meaningful to the members of a given society or group. The group members (of an organisation or culture) are the judges of the validity of an *emic* account. Again, for simplicity, call it the *subjective* account.

Now, the question arises: is one of these accounts "truer" than the other? Is one of them more fundamental or more reliable? Which account will best give us the information we need in order to understand what is going on in society?

Cultural materialism is a research orientation in social anthropology that makes the *etic-emic* distinction, but tends to see the *etic* as primary and the *emic* as a complementary (but relatively passive) explanatory. (Really extreme materialists – rare these days – see the *emic* context of mind and consciousness as an epiphenomenon – i.e. something that arises out of objective, physical world processes, but which has no actual, reciprocal causal effect on the world; i.e. "mind is an illusion"). The relative dominance of the *etic* account is understandable if your principle aim is to study and describe something "objectively", i.e. *without disturbing it.* However, for policy analysis and for decision making such an

[2]Pike introduced these terms in the context of linguistics, where they were applied in a specific way. However, they are now well established in the social sciences in a more general manner. That these concepts had to be re-introduced into Anglo-Saxon social science in the early 1950 is, for me, a mystery. It may be that they simply got lost in the shuffle, along with so much else, during the 1914–1945 period.

approach is deluded: you will never begin to understand a society (or any living system) until you start poking it with a sharp stick to see how it reacts.

Clearly, if your aim is not just to *study and describe* a social system (a population, an organisation or an institution), but to *do* something with it – i.e. to intervene, to change, to develop – then the *emic* account is not simply a complementary explanatory. It becomes an all important context for understanding what is actually going on. More correctly, it is the *interaction between* the *etic* (objective) and the *emic* (subjective) – as fully equal and *efficient causal contexts* – that drives the development of human social systems in an open-ended manner. And this interaction is characterised by something called *genuine uncertainty.*

3.3 Genuine Uncertainty (GU)

People often confuse the notions of *risk* and *uncertainty*. However, these are completely different animals. The person who is usually credited with establishing the distinction between these two concepts (at least in the field of economics) is Frank Knight in his work "Risk, Uncertainty, and Profit" (1921):

> ". . . Uncertainty must be taken in a sense radically distinct from the familiar notion of Risk, from which it has never been properly separated. . . . The essential fact is that "risk" means in some cases a quantity susceptible of measurement, while at other times it is something distinctly not of this character; and there are far-reaching and crucial differences in the bearings of the phenomena depending on which of the two is really present and operating. . . . It will appear that a *measurable* uncertainty, or "risk" proper, as we shall use the term, is so far different from an *un-measurable* one that it is not in effect an uncertainty at all". (Emphasis added).

Risk is defined as a type of uncertainty based on a well grounded (quantitative) probability. Formally, Risk = (the probability that some event will occur) x (the consequences if it does occur). Where probability is relevant, one can calculate risk. And if risk has a well grounded probability then, as Knight says, there is, in effect, no uncertainty at all.

Genuine uncertainty, on the other hand, embodies processes and outcomes which cannot be ascribed (well grounded) probabilities. Thus in the world of social, political, organisational and ideological systems (i.e. in working with "subjective problems"), what is often called *risk analysis* is not about *risk* at all, but is properly a matter of *genuine uncertainty.*

Needless to say, banks, insurance companies, stockbrokers and other serious gambling establishments love *risk* and hate *uncertainty*. This is exemplified by insurance companies essentially being willing to insure *anything* to which they can assign a well-grounded probability while avoiding anything that they cannot (ironically called "Acts of God").

It gets worse: besides not being amendable to quantification, processes involving GU do not necessarily have *specified outcome spaces.*

Suppose someone asks me: "What will the population of Sweden be in the year 2500?" (assuming that what we call "Sweden" is still around). My answer is: "I haven't got the foggiest!" It could be anything – almost. However, there is one thing I do know with great certainty: it will be a number between 0 and (let's say) 100 billion. That is to say, I don't know the number, but I know what the "outcome space" looks like. It is the counting manifold representing zero and the positive integers. The outcome space is fully *specified.*

Now, suppose someone asks me: "What new scientific discoveries will be made in the next 500 years?" Again, my answer is: "I don't have the foggiest!" But this time, I don't even know how to categorise the possibilities. I have no certain knowledge about the "outcome space" (i.e. where to *place* such *unknown unknowns* in relation to what we know today) since this space does not exist yet. This type of GU involves an open-ended set of possibilities, in which the "outcome space" is *unspecified.*

Unspecified uncertainty is especially relevant when working with long-term future developments. This type of uncertainty is inherently ineradicable – you cannot get rid of it by trying to obtain more information about it, because the information needed to reduce it simply isn't there (yet).

Finally there is the issue of so-called *agonistic uncertainty.* The word-stem *agon* comes from (classical) Greek and means a "contest" or (mental) "struggle" (compare the words *agony* and *antagonistic*). Agonistic uncertainty has to do with competing and cooperating *actors*, *wills* and *intensions* (think of the stock market). In terms of a *complex self-referential system,*[3] it refers to a network of conscious agents (e.g. individuals, organisations, institutions, nations) acting concurrently and reacting to each other. The behaviour of the system is *emergent* as opposed to predetermined (i.e. it produces surprises), and its development is unpredictable.

To sum up, the *genuine uncertainty* inherent in WPs is characterised by three (intertwined) properties that defy prediction:

- It cannot be ascribed a (well-grounded) probability (therefore you cannot predict the "odds" of certain things happening).
- It does not have a well-defined or complete outcome space, but is full of "unknown unknowns" and emergent processes (so you cannot even predict *what might* happen).
- It involves *subjective, self-referential* behaviour (which means that *meta-actors* can consciously decide to do unexpected, surprising things).

[3]One of the most well known books about this is Hofstadter: *Gödel, Escher, Bach: An Eternal Golden Braid* (1979). For a book more focused on self-reference in social systems, see Luhmann (1995).

3.4 Ten Criteria for Wicked Problems

It is instructive to look at the original criteria put forward by Rittel and Webber to characterise WPs. (It has been pointed out that some of these criteria are closely related or have a high degree of overlap, and that they should therefore be condensed into four or five more general criteria. I think that this is a mistake, and that we should treat these criteria as 10 heuristic perspectives which will help us better understand the nature of complex social planning problems.)

1. There is no definite formulation of a wicked problem.
   "The information needed to *understand* the problem depends upon one's idea for *solving* it. This is to say: in order to *describe* a wicked problem in sufficient detail, one has to develop an exhaustive inventory for all the conceivable solutions ahead of time". [This seemingly incredible criterion is in fact treatable. See below.]
2. Wicked problems have no stopping rules.
   In solving a tame problem, "... the problem-solver knows when he has done his job. There are criteria that tell when *the* solution or *a* solution has been found". With wicked problems you never come to a "final", "complete" or "fully correct" solution – since you have no objective criteria for such. The problem is continually evolving and mutating. You stop when you run out of resources, when a result is subjectively deemed "good enough" or when we feel "we've done what we can..."
3. Solutions to wicked problems are not true-or-false, but better or worse.
   The criteria for judging the validity of a "solution" to a wicked problem are strongly stakeholder dependent. However, the judgments of different stakeholders ..."are likely to differ widely to accord with their group or personal interests, their special value-sets, and their ideological predilections". Different stakeholders see different "solutions" as simply better or worse.
4. There is no immediate and no ultimate test of a solution to a wicked problem.
   "... any solution, after being implemented, will generate waves of consequences over an extended – virtually an unbounded – period of time. Moreover, the next day's consequences of the solution may yield utterly undesirable repercussions which outweigh the intended advantages or the advantages accomplished hitherto".
5. Every solution to a wicked problem is a "one-shot operation"; because there is no opportunity to learn by trial-and-error, every attempt counts significantly.
   "... *every* implemented solution is consequential. It leaves "traces" that cannot be undone ... And every attempt to reverse a decision or correct for the undesired consequences poses yet another set of wicked problems ...".
6. Wicked problems do not have an enumerable (or an exhaustively describable) set of potential solutions, nor is there a well-described set of permissible operations that may be incorporated into the plan.

"There are no criteria which enable one to prove that all the solutions to a wicked problem have been identified and considered. It may happen that no solution is found, owing to logical inconsistencies in the 'picture' of the problem".

7. Every wicked problem is essentially unique.
   "There are no *classes* of wicked problems in the sense that the principles of solution can be developed to fit *all* members of that class". . . .Also, . . ."Part of the art of dealing with wicked problems is the art of not knowing too early which type of solution to apply".
8. Every wicked problem can be considered to be a symptom of another [wicked] problem.
   Also, many internal aspects of a wicked problem can be considered to be symptoms of other internal aspects of the same problem. A good deal of mutual and circular causality is involved, and the problem has many causal levels to consider. Complex judgements are required in order to determine an appropriate *level of abstraction* needed to define the problem.
9. The causes of a wicked problem can be explained in numerous ways. The choice of explanation determines the nature of the problem's resolution.
   "There is no rule or procedure to determine the 'correct' explanation or combination of [explanations for a wicked problem]. The reason is that in dealing with wicked problems there are several more ways of refuting a hypothesis than there are permissible in the [e.g. physical] sciences".
10. [With wicked problems,] the planner has no right to be wrong.
   In "hard" science, the researcher is allowed to make hypotheses that are later refuted. Indeed, it is just such hypothesis generation that is a primary motive force behind scientific development (Ritchey, 1991). Thus one is not penalised for making hypothesis that turn out to be wrong. "In the world of . . . wicked problems no such immunity is tolerated. Here the aim is not to find the truth, but to improve some characteristic of the world where people live. Planners are liable for the consequences of the actions they generate . . ."

3.5 Tackling Wicked Problems with General Morphological Analysis

How, then, does one tackle wicked problems? Some 20 years after Rittel & Webber wrote their article, Rosenhead (1996), of the London School of Economics, presented the following criteria for dealing with complex social planning problems – criteria were clearly influenced by the ideas presented by Rittle, Webber and Ackoff:

- Accommodate multiple alternative perspectives rather than prescribe single solutions
- Function through group interaction and iteration rather than back office calculations

- Generate ownership of the problem formulation through transparency
- Facilitate a graphical (visual) representation for the systematic, group exploration of a solution space
- Focus on relationships between discrete alternatives rather than continuous variables
- Concentrate on possibility rather than probability

As I will attempt to argue in the coming sections of this book, group facilitated GMA is fully attuned to these criteria. As a preview, let us take some of Rittel &Webber's criteria for WPs and see how GMA stacks up.

Criterion #1. *... in order to describe a wicked problem in sufficient detail, one has to develop an exhaustive inventory for all the conceivable solutions ahead of time.*

Done properly, GMA results in an inference model which strives to represent the total problem space, and as many of the *potential solutions* to the given problem complex as possible. This goes a long way in satisfying this seemingly incredible criterion. The idea is to "play" with the inference model in order to allow stakeholders to better understand the problem space and the possible consequences of alternative decisions. We literally build up an inventory of all possible solutions, before we begin to fully understand the problem.

Criterion #3. *Different stakeholders ..."are likely to differ widely to accord with their group or personal interests, their special value-sets, and their ideological predilections".*

The *process* of creating morphological inference models through facilitated group workshops is as important as the end-product – i.e. the model itself. As many stakeholders as possible should be engaged in the work, in order to create a common terminology, common problem concept and common modelling framework. Principal stakeholders and subject specialists should therefore be brought together in a series of workshops to collectively (1) structure as much of the problem space as possible, (2) synthesize solution spaces, (3) explore multiple solutions on the basis of different drivers and interests and (4) analyse stakeholder structures. The different stakeholders do not have to agree on a single, common solution, but must be encouraged to *understand each other's positions and contexts.*

This last point is crucial. *Consensus* means "general agreement or concord" within a group. Facilitators usually differentiate between *first-order* and *second-order* consensus. The normal *first-order* form is that of gaining a common standpoint or agreeing upon a common solution. This is seldom the case with stakeholder groups working with wicked problems. So-called *second order consensus* is when stakeholders in a group learn to accept each other's specific stakeholder positions – on the basis of understanding the *reasons* for these positions. (This is called "position analysis" in Swedish, and is a discipline in itself.)

Criterion #7. *...part of the art of dealing with wicked problems is the art of not knowing too early which type of solution to apply.*

In GMA we call this "remaining in the mess", i.e. keeping one's options open long enough to explore as many relationships in the problem topology as possible, before starting to formulate solutions. This can be a frustrating process for inveterate "problem solvers", but is an absolutely necessary procedure when modelling wicked problems.

Criterion #8. *Every wicked problem can be considered to be a symptom of another [wicked] problem.*

With a morphological inference model, one can treat any particular parameter or "issue" as the starting point or "independent" variable. This allows one to change perspectives and treat different issues as both causes (drivers) and effects (or symptoms). Everything is connected, which is what both wicked problems and GMA is about.

Criterion #10. *[With wicked problems,] the planner has no right to be wrong.*

Not only should planners be part of the GMA modelling and shaping process, but also the potential "consumers" or "victims" of said planning. People do not like to be "planned at", without having something to say about it. Representative "consumers" of the planning project must absolutely be made part of the planning process itself. GMA allows for – almost insists upon – this type of participation.

Finally, as is the case with all modelling methods dealing with complex social planning problems, there is always the *garbage in–garbage out* problem. However, even here group facilitated GMA has some clear advantages. It expressly provides for a good deal of in-built "garbage detection", since poorly defined parameters and incomplete ranges of conditions are immediately revealed when one begins the task of cross-consistency assessment. These assessments simply cannot be made until the morphological field is well defined and the working group is in agreement about what these definitions mean. This type of garbage detection is extremely important when working with *wicked problems* and *social messes*.

Chapter 4
Modelling Complex Policy Issues with Morphological Analysis

GMA is especially suitable for pitting one type of structure against another – for instance a scenario model against a strategy model. We call this *duplex* modelling – in contrast to *simplex* modelling.

We have borrowed these terms from geometry and topology. There the term *simplex* is used to denote basic geometric elements (a line, triangle or other geometric figure), which can be multi-dimensional, but which are "simply" connected. What we mean by a *simplex model* in the context of GMA is that the model consists of a number of dimensions which are all in the *same context* or of the same *conformity*, so to speak. We use the term "duplex" or "multiplex" to denote models that consist of different field segments or parameter groups, which represent different contexts or frameworks. Examples of simplex models are:

- A scenario model (i.e. a scenario generation laboratory)
- A strategy model
- A stakeholder analysis
- An organisational structure

Duplex models pit a set of parameters representing one context against another set representing another context. For instance:

- Scenarios vs. strategies
- Strategies vs. organisational structure
- Stakeholders vs. strategies

The idea behind duplex (or multiplex) models is that we can treat one of the modelling contexts as input (or as starting conditions) and another as output (or as outcome conditions). Indeed, we can treat any context, or any subset of parameters within a context, as input, and examine any others as output. We will give an example of this below.

Portions of the text in this chapter are based on the articles Stenström and Ritchey (2004); Ritchey (2006a).

T. Ritchey, *Wicked Problems – Social Messes*, Risk, Governance and Society 17,
DOI 10.1007/978-3-642-19653-9_4, © Springer-Verlag Berlin Heidelberg 2011

Concerning social and organisational modelling, there is a more or less established way of classifying the different principal contexts or environments which can be modelled. These are:

- The contextual environment
- The transactional environment
- The policy or strategy space

The *contextual environment* is made up of those factors in the *external world* which can influence how organisations (or systems) function, but which cannot be (practically) influenced by (most) organisations. For instance: the weather, technological breakthroughs, oil prices, legal structures.

The *strategy space* is defined as the internal world of an organisation or system, comprising those factors which the system-owner *can* control and mould into a strategy for coping with the contextual environment. New product development, marketing strategies, R&D directives and organisational structure are examples.

However, what is to be designated as "external" and "internal" is a practical matter and may vary depending on the purpose of a study. In reality, there is always some degree of overlap between these contexts. Some factors, while being external to an organisation as such, *may* be influenced by deliberate actions carried out by the organisation. For instance, a population's inclination to buy a certain type of product is not part of an organisation's (internal) strategy space, but may well be influenced to some degree by the choice of strategy (e.g. information campaigns, rewards or sanctions) (Fig. 4.1).

Factors, which are external to an organisation or system as such, but which can be influenced by that organisation/system, belong to the so-called *transactional environment*.

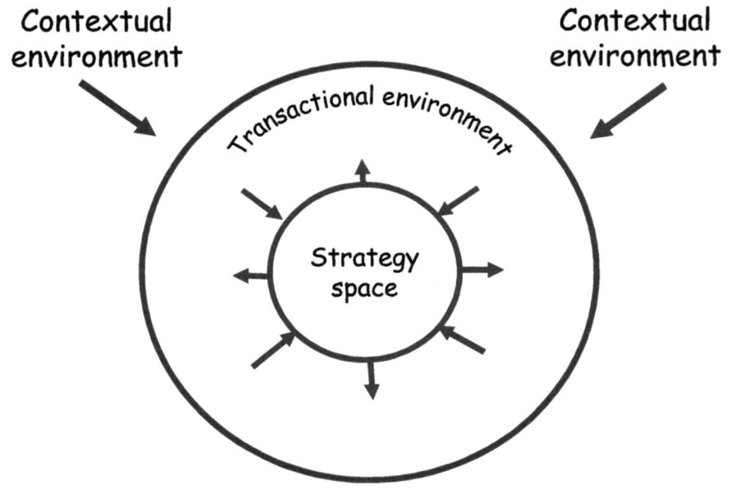

Fig. 4.1 Three modelling environments

GMA-projects typically involve developing duplex models which allow us to pit a contextual environment (or "external scenario field") against an organisational "strategy space". Such *conceptual laboratories* have been developed as instruments for evaluating, e.g. antagonistic threats vs. alternative preparedness strategies; international developments vs. national security policies; or climate scenarios vs. new insurance instruments.

We have found GMA especially suitable in the area of environmental policy studies, in which different environmental strategies can be pitted against a range of possible futures projections. A case in point was a study done for the Swedish Ministry of Environment concerning the development of an *Extended Producer Responsibility* system in Sweden (Stenström & Ritchey, 2004).

Extended producer responsibility (EPR) imposes accountability over the entire life cycle of products and packaging introduced on the market. This means that firms, which manufacture, import and/or sell products and packaging, are required to be financially or physically responsible for such products after their useful life cycle. They must either take back spent products and manage them through reuse, recycling or in energy production, or delegate this responsibility to a third party – a so-called producer responsibility organization (PRO), which is paid by the producer for spent-product management. In this way, EPR shifts responsibility for waste from government to private industry, obliging producers, importers and sellers to internalise waste management costs in their product prices (see Hanisch, 2000).

The long-term purpose of EPR is to encourage more environmentally friendly product development – products that require fewer resources, which are easier to reuse/recycle, and which contain fewer environmentally dangerous substances. The problem, then, is to develop flexible EPR-strategies for a future in which there is a good deal of uncertainty concerning, for instance, national and international directives, technological developments, shifting political ideologies, market forces and ethical concerns. The purpose of the EPR morphology was to systematically formulate a range of future contextual environments by which to test alternative EPR strategies.

Two complementary morphological models were developed: one with which we could systematically formulate and explore different possible *futures projections* based on factors which cannot be directly controlled by the organisations responsible for developing the EPR-system (i.e. an "external world" field or contextual environment); and one with which we could systematically formulate and explore different EPR *strategies,* depending on variables which can be more or less controlled (i.e. an "internal world" field or strategy space). These models were then linked to each other by cross-consistency assessments, in order to establish which strategies would be most effective and flexible for different ranges of scenarios.

For this purpose, two working groups of seven persons each – a *contextual environment group* and a *strategy development group* – performed the modelling together with two morphologists. The groups were composed of researchers from the Swedish EPA and other relevant government authorities, from two NGOs and from two private companies involved in waste management and recycling. Each group worked for 2 days on their respective models, with a final 1-day joint session where the contextual environment model was merged with the strategy model.

SCENARIO	Buyer behaviour	Consumption patterns Total: Private import	Consumer sorting behaviour (trends)	National environmental policy	Price of new raw material vs reclaimed material	Production technology: volume of materials	Technology development: reclaiming technology	EU-directives for import and export of waste
Global Crisis (Production gone wild)	Willing to pay more for green products	Total: Up Private import: Up	Voluntary (ideologically driven)	At the forefront; Holistic approach (legal & econ.)	New: High Reclaimed: High	Much less than today	Very rapid increases	Less restricted than today
Raw Material Depletion	Will to buy green, but will not pay more	Total: Status Quo Private import: Up	Will sort for compensation/ reward	At forefront, but no holistic approach (legal only)	New: High Reclaimed: Low	Somewhat less than today	Substantial increases	Same as today
Current policies (Negative trend)	No interest in buying green products	Total: Up Private import: SQ	Will sort if facing sanctions	Ideological, based on voluntary acceptance	New: Low Reclaimed: High	Same as today	Only marginal increases	More restrictive than today
Current policies (Positive trend)		Total: SQ Private import: SQ	Will resist sorting	Least possible adaptation	New: Low Reclaimed: Low			
Green-house effect (Stop emissions)								
Batman: High-tech solutions								
Dematerialised production (New materials)								
Green market (ideological paradise)								

Fig. 4.2 An eight-parameter scenario field – one possible futures projection highlighted

Figure 4.2 shows an EPR *scenario field* consisting of eight parameters which represent "external" factors that can influence or constrain a Swedish EPR system. The eight parameters generate 20,736 formal configurations. In contrast to strategy fields, or fields representing system solutions, scenario fields are often difficult to assess internally and reduce. This is because it is risky to exclude relationships which may seem improbable today, but which might very well be the case in 5, 10 or 50 years. In such cases, it is better to work backwards, so to speak: Select one or more parameters as drivers, choose a number of configurations based on varying these drivers, and then assess the chosen configurations for internal consistency. Repeat this process until the desired number of scenario projections is achieved.

For the study in question, eight *specific configurations* were chosen. Together, these covered all of the parameter states in the scenario field and represented a broad range of futures projections. The configurations were then named and linked to the column at the far left, a scenario-name "placeholder".

This is done for practical reasons, in order to keep track of specific configurations of interest. (When such a placeholder is employed to define specific configurations, we call the field *specified*. When no such placeholder is present, then the field is *open*.)

Note: On the computer, morphological field configurations are colour-coded. For instance, selected input conditions are rendered in red, and output conditions in blue. In the figures below, red is represented by grey, and blue is represented by black.

EPR rules and regulations	Environmental adaptation of products	Required range of information about products	Waste sorting system	Collection system	Recycling system	Dominant EPR market for waste products	Instruments for deposition and burning
Voluntary, branch regulated	Focus on clean materials	Chemicals Material Energy	> 15 commodity groups	Very near premises	Mechanical recycling	International	Recycling: Up Energy: Down
General legislation toward individual. No monopoly.	Same mix as today	Chemicals Material	> 15 material groups	High density "bring system"	Thermal recycling	National and close international	Recycling: Up Energy: Up
General legislation toward collective Partial monopoly.	Focus on dematerialisation	Chemicals Energy	Same as today	Low density "bring system"	Chemical recycling	Local/regional	Recycling: Down Energy: Up
Finely detailed legislation (who, how & what)		Chemicals only	< 5 commodity groups		Biological recycling		Relative increase of deposition
			< 5 material groups				

Fig. 4.3 An eight-parameter strategy field – one possible strategy highlighted

Figure 4.3 is a *strategy field* which also (purely coincidently) contains eight parameters. It represents important "internal factors" of a (future) Swedish EPR system. The field generates 34,560 formal (strategy) configurations. A cross-consistency assessment reduced this to 480 strategies which were deemed realistic. An explicate strategy placeholder parameter was not employed with this field, since we wished it to be left "open". The reason for this will be made clear below.

The scenario and strategy fields can be linked in order to test the viability of different strategies against chosen futures projections. However, fully linking these two eight-parameter fields into a 16 parameter field would result in a combined field consisting of over 700 million formal configurations. Although there are no purely technical constraints to working with such a large field, it produces an intimidatingly large cross-consistency matrix. Fortunately, we can get around this problem by using a condensed form of the scenarios: we simply merge the scenario "placeholder" parameter with the strategy field (see Fig. 4.4).

There are two ways to make the cross-consistency assessments between the scenario placeholder parameter and the strategy parameters – a *quick method* and a *thorough method*. The quick method involves relating each scenario, *as a gestalt*, to each of the strategy parameters. The group making these assessments should, of course, refer to the complete scenario field, but only in order to form a *total picture* of what each scenario configuration would imply for *each state* of each strategy parameter. There is no direct assessment between the *internal states* of a scenario and the strategy parameters. This quick method is usually employed when there is limited time for group work.

SCENARIO	EPR rules and regulations	Environmental adaptation of products	Required range of information about products	Waste sorting system	Collection system	Recycling system	Dominant EPR market for waste products	Instruments for deposition and burning
Global Crisis (Production gone wild)	Voluntary, branch regulated	Focus on clean materials	Chemicals Material Energy	> 15 commodity groups	Very near premises	Mechanical recycling	International	Recycling: Up Energy: Down
Raw Material Depletion	General legislation toward individual. No monopoly.	Same mix as today	Chemicals Material	> 15 material groups	High density "bring system"	Thermal recycling	National and close international	Recycling: Up Energy: Up
Current policies (Negative trend)	General legislation toward collective Partial monopoly.	Focus on dematerialisation	Chemicals Energy	Same as today	Low density "bring system"	Chemical recycling	Local/regional	Recycling: Down Energy: Up
Current policies (Positive trend)	Finely detailed legislation (who, how & what)		Chemicals only	< 5 commodity groups		Biological recycling		Relative increase of deposition
Green-house effect (Stop emissions)				< 5 material groups				
Batman: High-tech solutions								
Dematerialised production (New materials)								
Green market (ideological paradise)								

Fig. 4.4 The scenario-name placeholder parameter (at *far left*) imposed on the eight-parameter strategy field. One possible scenario–strategy configuration highlighted

SCENARIO	EPR rules and regulations	Environmental adaptation of products	Required range of information about products	Waste sorting system	Collection system	Recycling system	Dominant EPR market for waste products	Instruments for deposition and burning
Global Crisis (Production gone wild)	Voluntary, branch regulated	Focus on clean materials	Chemicals Material Energy	> 15 commodity groups	Very near premises	Mechanical recycling	International	Recycling: Up Energy: Down
Raw Material Depletion	General legislation toward individual. No monopoly.	Same mix as today	Chemicals Material	> 15 material groups	High density "bring system"	Thermal recycling	National and close international	Recycling: Up Energy: Up
Current policies (Negative trend)	General legislation toward collective Partial monopoly.	Focus on dematerialisation	Chemicals Energy	Same as today	Low density "bring system"	Chemical recycling	Local/regional	Recycling: Down Energy: Up
Current policies (Positive trend)	Finely detailed legislation (who, how & what)		Chemicals only	< 5 commodity groups		Biological recycling		Relative increase of deposition
Green-house effect (Stop emissions)				< 5 material groups				
Batman: High-tech solutions								
Dematerialised production (New materials)								
Green market (ideological paradise)								

Fig. 4.5 Scenario–strategy matrix with three inputs (*grey*)

The *thorough method* assesses the relationships between the internal states of each (defined) scenario configuration, and the internal states of each of the strategy parameters. This requires 8 times as many evaluations (since, in this case, there are eight internal elements for each scenario configuration), but is it much more rigorous and provides an interesting base for discussions (a crucial aspect of all the phases of a morphological analysis).

In working with linked morphological fields, there are no automatically designated independent variables or drivers. Any parameter – or set of parameters – can be designated as such. Thus anything can be designated input and anything output. For instance, instead of simply letting a scenario placeholder define a relevant strategy, one can reverse the process and allow chosen states within a proposed strategy to designate relevant scenarios.

Figure 4.5 is an example. In this case, we have essentially posited the following question to the model: "If we want to develop an EPR system based on *general legislation* and *international markets*, with emphasis on detailed *material-group sorting*, what are the other consistent (internal) conditions for such a system, and with which (external) scenario configurations is this system most compatible?" This feature, of being able to define any combination of conditions as inputs – even mixing external and internal conditions – gives morphological models great flexibility.

Chapter 5
Strengths, Limitations and Advanced Topics

As with all problem structuring and modelling methods, GMA has its strengths and its limitations. These will be outlined below. GMA can also be utilised in conjunction with other modelling procedures, usually as an initial step in framing "unstructured reality" – i.e. to create a comprehensible *first modelling framework* when faced with a social/organisation mess.

Finally, we will outline a number of extended features that have been developed in order to enhance the functionality of GMA and expand its areas of application. It was only with the introduction of dedicated computer support some 15 years ago that has made it possible to develop these features.

5.1 GMA's Strengths

GMA straddles the fence between "hard" and "soft" scientific modelling. It is built upon the basic scientific method of going through cycles of analysis and synthesis and "parameterizing" a problem space. It defines structured variables and thus creates a *real* model, so to speak, in the form of a linked variable space in which inputs can be given, outputs obtained, and hypotheses ("what-if" assertions) made. For this reason, MA is compatible with other modelling procedures, and can be employed as a test-bed or first step in the development of other types of models (see below).

GMA also has definite advantages for group discussions and modelling work. As a process, the method demands that the problem space be clearly defined. Poorly defined concepts become immediately evident when they are cross-referenced and assessed for internal consistency. In this sense, GMA's cross-consistency assessment (CCA) acts both as a "garbage detector" and an effective means in ironing out vague concepts and terminological differences.

This last point should not be underestimated. I am constantly amazed by the fact that a group of subject matter specialists – who may have been working in the same

Text in this chapter is based on earlier published articles on GMA, including Ritchey, (1991, 2005b) and De Waal and Ritchey (2007).

T. Ritchey, *Wicked Problems – Social Messes*, Risk, Governance and Society 17,
DOI 10.1007/978-3-642-19653-9_5,

general area for 20 years or more (for instance transport security, environmental threats or crime mitigation) – when bought together can have vastly difference experiences, perspectives and terminologies. These differences – which can make it difficult for them even to agree upon what the "real problem" *is* – depend on their specific specialities or (more importantly) with which "stakeholder" groups they are associated. In this context, one of the most important results of a GMA workshop is creating among the participants a common conceptual framework, terminology and understanding. This creates "smart groups" or "smart teams".

Finally, GMA leaves an acceptable *audit trail.* One of the main problems in working with *wicked problems* is that the actual process by which they are formulated and structured is often difficult to trace – i.e. we seldom have an adequate *audit trail* describing the process of getting from underlying assumptions, to initial problem formulation, to specific solutions or conclusions. Without some form of *traceability* we have little possibility of scientific control over results, let alone reproducibility. The software supported documentation of every concept and each and every cross-consistency assessment in a morphological model creates such an audit trail.

5.2 GMA's Limitations

"*GMA requires strong, experienced facilitation* (if this is to be considered a limitation: see Chap. 7, Facilitating GMA Workshops). Parameterizing a problem space" by creating and linking structured variables is considerably more difficult and time consuming than developing an influence diagram containing "black box" variables and associated "arrows". Without proper facilitation, it is very easy to create trivial morphological models.

GMA takes time. Meaningful morphological models cannot be created in an afternoon. Depending on the complexity of the problem and the level of ambition, developing a morphological model can take between 2 and 6 group-workshop days. The work described in Chap. 4, concerning the development of a model for Extended Producer Responsibility strategies, took 5 workshop days. We have done studies which have required up to 20 workshop days under an 18 month period.

GMA cannot be effectively carried out in groups larger than 7–8 participants, where the whole point is to foster dialog between subject specialists. The threshold of group dynamics, which separates participants talking to one another, from participants addressing a group as a whole, is astonishingly consistent at the magic number 7 ± 2.

Proper morphological modelling requires dedicated computer support. Doing group work with the type of problems described in this book is virtually impossible without such support. This is why GMA is only now developing into its full potential.

Finally, as with all modelling methods, the outcome of a morphological analysis is no better than the quality of its input. It is the responsibility of the facilitator – in collaboration with the client – to make sure that a competent group is formed, and that the GMA modelling process is carried out properly.

5.3 Use in Combination with Other Methods

Since the central feature of morphological analysis is to parameterize a problem complex and examine its internal coherence, GMA can be used to good advantage as a *preceding step* which can provide input for the development of other modelling techniques. The connection between GMA and Bayesian Network (BN) modelling is an especially interesting example.

A BN is a graphical structure (technically a *diagonal acyclic graph* or DAG) representing cause–effect relationships between a number of defined variables. Each variable is assigned a range of mutually exclusive *values* or *states*, and the causal relationships between them are quantified by means of probabilities. Once a BN is quantified, it can propagate newly acquired information through the rest of the network.

GMA and BN are thus closely related methods for developing inference models. Each has its advantages and disadvantages for modelling complex processes and systems. MA allows small groups of subject specialists to define, link and internally evaluate the parameters of complex problem spaces, thus creating a solution space and a flexible inference model. However, GMA cannot easily deal with hierarchical structures and causal relationships.

Bayesian Networks allow for such causal and hierarchal relationships, but they are more difficult to employ in the initial, problem formulation phase of the modelling process. Combining GMA and BN, as two phases in a modelling process, allows us gain the benefits of both of these modelling methods.

When constructing a BN model, the major modelling criteria that arise are:

1. What are the relevant, principal variables of the problem complex?
2. What are the values ranges of the different variables?
3. Between which of the variables are there dependencies?
4. What are the causal directions of these dependencies?
5. What are the strengths of the dependencies, as depicted in the graphical structure?

As can immediately be seen, the first three steps in this process are realized in a morphological analysis.

In earlier work with Bayesian Networks, we have found that the very prospect of tackling all of the five modelling steps from scratch, under limited time conditions, was truly daunting for the working group. We also found a tendency to rush into (or to pre-suppose conditions in) steps 4 and 5 before steps 1 and 2 of the process were mature enough, causing a good deal of confusion. The whole process becomes much more tenable if it is broken up into *two conceptually distinct modelling phases*: do a GMA first, without any reference to directed causality or hierarchy, thus allowing the working group to concentrate on one main task. When this is accomplished, steps 4 and 5 in the BN modelling process follow much more easily.

Another modelling method, which can be supported by GMA, is multi-criteria decision analysis. Particular solutions coming out of a morphological model, whether these are scenarios, strategies, or other types of configurations, can be employed as *alternatives* in, e.g. the Analytic Hierarchy Process (AHP) (Saaty, 2001). AHP is a

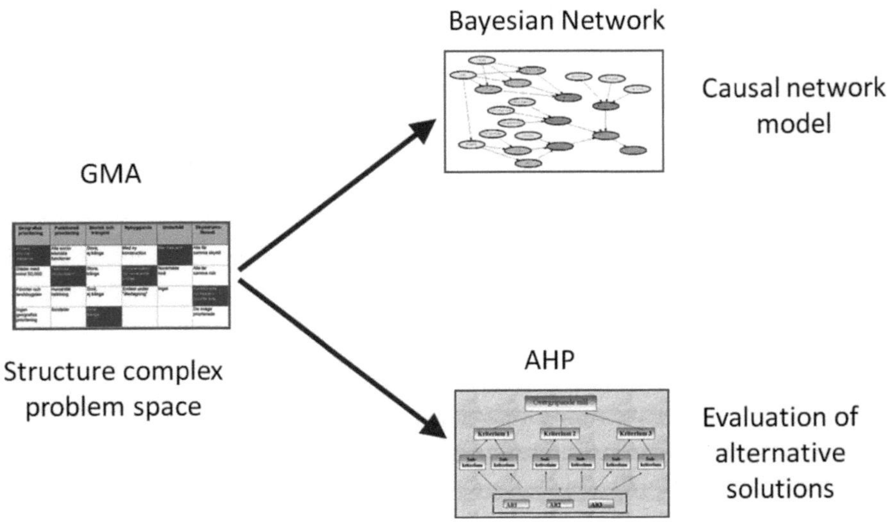

Fig. 5.1 GMA as a precursor to BN and AHP

method for systematically comparing alternative solutions in the context of a hierarchy of goals and goal criteria (Fig. 5.1).

5.4 Advanced Topics

5.4.1 *Multi-Part Internal Evaluations*

Cross-consistency assessments done on a morphological field treat of *pair-wise relationships* between parameter values. These assessments are carried out, *inter alia*, in order to identify inconsistent conditions in the parameter space, thus reducing this space and defining a solution space. However, it often happens that a pair of parameter values is consistent or inconsistent *depending on the value of a third parameter*. In many cases this causes no problem: if a pair-wise relationship is possible under *any circumstances*, then it is possible, and should not be forbidden. However, in some instances it is important for the model to explicitly account for this particular *conditional dependency*. An example is the "cumulative inconsistency" of increasing costs or other quantities that may be represented *across several parameters* at once. The modelling system allows for treating multi-part parameter assessments.

However, if such multi-part assessments are a dominant feature of the model, then this is a sign that one should at least explore the possibility of further developing the morphological model into a Bayesian Network model.

5.4.2 AND-Lists

Strictly speaking, a *true variable* consists of mutually exclusive values or states. However, it is sometimes advantageous to formulate a parameter consisting of values or states which are *not* mutually exclusive. Variables containing mutually exclusive values are called "OR-lists" (i.e. their logical relations are based on the Boolean "or" operator). Variables containing values which *can exist concurrently* are called "AND-lists". Each of the values in an AND-list can be thought of as a simple binary variable: for every other parameter value in the model P_x, it is either "on" (i.e. compatible with P_x) or "off" (incompatible with P_x). AND-lists are useful for saving parameter space and condensing many simple "yes–no" variables into a single parameter. They are best used as output parameters, expressing e.g. concurrent goals or methods. However, they can also be employed in other ways.

5.4.3 Stakeholder or Position Analysis

Sets of AND-lists can be employed in a stakeholder analysis. This can be done by first formulating a conventional morphological field, such as the EPR strategy field in Chap. 4, and then letting different stakeholders define their respective "positions" for each of the parameters in the field. A new field is then created by treating each stakeholder as a parameter, with the stakeholder *positions* concerning the strategy field listed beneath (see Fig. 5.2). The group of stakeholders then does a cross consistency assessment on this field. This is an extremely interesting exercise which can be applied to negotiations.

5.4.4 Time Lines

Time can be treated in a number of ways in morphological models:

- *Naked time parameter*: This is a parameter which simply lists time intervals as such (e.g. within an hour, within a day, within a week, etc.). Any other parameter, which is dependent on time, can then be related to this general time parameter. It can then be used as a co-driver with any other driver or drivers, in order to examine a time-line or critical time points.
- *Applied time parameter*: This is a parameter which measures a time line for a *specific process or event*. Any other parameter, which is dependent upon this process or event, can then be related to it.
- *Parameter-wise time ordering*: In this case, some or all of the parameters in the field are ordered (left to right) in a time-line. This is used when the order of the parameters represents a time-ordered process.

Stakeholder 1	Stakeholder 2	Stakeholder 3	Stakeholder 4
Stakeholder's position on parameter 1	Stakeholder's position on parameter 1	Stakeholder's position on parameter 1	Stakeholder's position on parameter 1
Stakeholder's position on parameter 2	Stakeholder's position on parameter 2	Stakeholder's position on parameter 2	Stakeholder's position on parameter 2
Stakeholder's position on parameter 3	Stakeholder's position on parameter 3	Stakeholder's position on parameter 3	Stakeholder's position on parameter 3
Stakeholder's position on parameter 4	Stakeholder's position on parameter 4	Stakeholder's position on parameter 4	Stakeholder's position on parameter 4
Stakeholder's position on parameter 5	Stakeholder's position on parameter 5	Stakeholder's position on parameter 5	Stakeholder's position on parameter 5

Fig. 5.2 Replica stakeholder field consisting of four AND-lists

Micro scenario (Time-line)	Information assurance	What command level required	Demands on quality of advice/decisions	SKI's required cooperation with others	What output required from SKI	Primary receiver of output from SKI
1. Sudden shift in threat perception	High assurance Almost real time	SKI & SSI together	Well analyzed, established advisory decision	Centralized, authority cooperation	Decisions concerning others' operations	Central government administration/PM
2. Observation of unauthorized activity outside of facilty perimeter	High assurance Short delay	SSI alone	Intensified expert analysis	Ski at location in Sweden	Expert advice by external demand	Central government authorities
3. Encroachment of perimeter observed	High assurance Long delay	SKI alone	Standard expert analysis	Assistance abroad	Descriptive information by external demand	Affected county administation
4. Encroachment of vital area of inner facility	Uncertain Almost real time	Watch commander	Simple analysis with expert support	Cooperation at a distance	Expert advice at own discretion	Police
5. Aggressor take control of vital facility area	Uncertain Short delay		Real-time deliberation	None	Descriptive information at own discretion	Municipal rescue services
6. Threat is made	Uncertain Long delay				None required	SSI
7. Demands made						SKI's line organization
8. Negotiation						Directly affected organization
9. Threat is carried out						

Fig. 5.3 Scenario field consisting of a scenario time-line (*far left* parameter) which steps through a series of configurations – one of which is highlighted

- *Configuration-wise time ordering*: This is a sequence of configurations which represents time-ordered development. It is especially useful for developing time-lines in scenario models (see Fig. 5.3).

5.4.5 Relational Database Applications

The documentation entered into the text areas associated with each cross consistency assessment can be collated into a relational database, which can then be addressed by defining drivers and configurations. This is useful when a lot of *structured information* is required in order to support a study.

5.4.6 Linking Morphological Fields

It is possible to allow the designated output of one morphological field to become the input for another field. Alternatively, the designated output a number of (sub-) fields can be collated into a single (super-) field. This allows for a hierarchical or networked morphological model. This is useful when a model treats of several levels of abstraction.

Chapter 6
On the Formal Properties of Morphological Models

6.1 Introduction

When we create models, whether these be quantified or non-quantified, we construct them on a scaffolding of *dimensions* – i.e. mental constructs which support a range of values or conditions. Together, these dimensions define a conceptual space. Spaces have certain properties, including metric relationships (geometry) and relationships of connectivity (topology).

The relationships within a conceptual space are dependent upon the nature of the concepts involved in *defining* the space. This was one of the main points of Bernhard Riemann's famous Habilitation paper "On the hypotheses which lay at the basis of geometry": i.e. that even the seemingly self-evident, three-dimensional physical space we live in is not given a priori, but is determined by its *content*, i.e. "in the forces which bind together its elements." (Riemann, 2004).

In the case of morphological spaces (or models) this is self-evident. Here dimensionality is not expressed in the form of continuous variables (as we do with physical space) but in the form of variables with well-defined, finite, discrete value ranges. And, in analogy to what Riemann points out concerning any theory of abstract spaces: in a discrete manifold, the principles of its *internal relations* are already given – or implied – by the specification of the dimensions of the manifold.

It is these "binding forces" – i.e. how the value ranges of a morphological model's dimensions relate to each other *logically, empirically and normatively for real world problems* – which make up the all important *content* of the model. In comparison, the purely *combinatorial* properties of morphological models are only the formal effects of this content.

However, the question does arise: is there any meaningful relationship between the formal properties of a morphological model and its empirical contents. This is

My thanks to Johan Schubert, mathematician/operational analyst at FOI, for scrutinizing this chapter, cleaning up the mathematical notation and providing valuable suggestions. All remaining faults are my own.

T. Ritchey, *Wicked Problems – Social Messes*, Risk, Governance and Society 17,
DOI 10.1007/978-3-642-19653-9_6, © Springer-Verlag Berlin Heidelberg 2011

a valid question which, we feel, warrants the study of these formal properties – on at least two grounds:

Firstly, if it were possible to classify morphological models into different "types" on the basis of purely formal characteristics, this might help us to better understand morphological modelling in general.

Secondly, there is a matter of pure academic curiosity. It would be intriguing to see to what extent morphological models/spaces *can*, in fact, be ascribed metric and topological properties *analogous* to those of the abstract spaces originally discussed by Riemann and, more recently, to the conceptual spaces ("geometry of cognitive representations") studied by Gärdenfors (2004).

This chapter is divided up into the following sections:

Firstly, we will look at some of the purely *combinatorial aspects* of morphological models, such as how the number of dimensions and the number of parameter values determines the size of the problem space and the cross-consistency matrix.

Then we will take a look at a number of *relationships* which are dependent upon the empirical content of the model, namely: how the *global connections between parameters* ("topology") and the internal relationships between parameters pairs ("geometry") influence the size and form of the solution space of the model.

Finally, we have selected a number of models and compared them on the basis of these derived relationships, in order to see what, if anything, it says about the nature of the models.

6.2 The Morphological Field

Let N = number of parameters in a morphological field (in Fig. 6.1, the Reference field, N = 5) and let P denote a Parameter such that the parameters in a morphological field are:

P1	P2	P3	P4	P5
P_1V_1	P_2V_1	P_3V_1	P_4V_1	P_5V_1
P_1V_2	P_2V_2	P_3V_2	P_4V_2	P_5V_2
P_1V_3	P_2V_3	P_3V_3	P_4V_3	P_5V_3
P_1V_4		P_3V_4		P_5V_4

Fig. 6.1 Reference morphological field

$$P_1, P_2, P_3 \ldots P_N.$$

Let v_x = the number of conditions in the value range of a given parameter P_x, such that the total morphological field is (quantitatively) defined by:

$$\{P_x v_i\}_{x,i}.$$

Then, the total number of *simple configurations* T_{SC} (i.e. a configuration with *one and only one condition designated under each parameter*) in a morphological field is:

$$T_{SC} = v_1 * v_2 * v_3 \ldots v_N$$

or

$$T_{SC} = \prod_{i=1}^{n} vi.$$

This simply shows that T_{SC} increases in a *factorial* manner ("geometrically" as it is sometimes referred to) with the increase in the number of parameters "n".

So much for the basic morphological field.

6.3 The Cross-Consistency Matrix and Parameter Blocks

The Cross-consistency matrix (CCM) pairs off every condition in each parameter with every other condition in all the other parameters. A *parameter block* (PB) consists of all of the paired conditions between two parameters, cross-referenced in the form of a 2-dimensional typology. In Fig. 6.2, the parameter blocks are shown in alternating shaded and white groups.

If N = number of parameters in a morphological field, then the number Parameter Blocks in the field's Cross-Consistency Matrix is:

$$\frac{1}{2}N(N-1).$$

This is an interesting mathematical expression that pops up all over that place. For instance:

- It is the formula for generating the triangular number series.
- It is the number of (non-directed) edges connecting N nodes in a graph.
- It is taught in social network theory and in facilitator training as the number of communication channels or possible (two-person) dialogues between

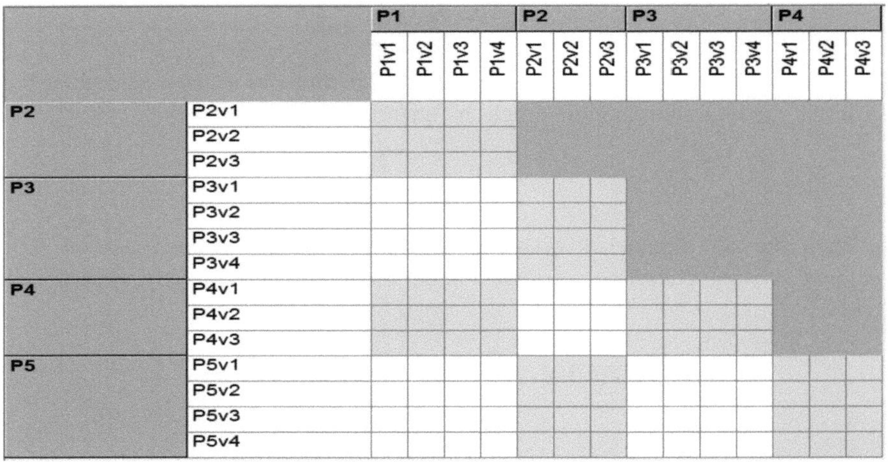

Fig. 6.2 Cross-consistency matrix (CCM) for morphological field in Fig. 6.1

N participants in a workshop (which is why group dynamics changes dramatically at around seven to eight people).

Of course, all this has a common base: Generally, ½N(N−1) is the number of dyadic (pair-wise) relationships between N elements or objects. It is equal to the binomial coefficient:

$$ {}_nC_k \equiv \frac{n!}{(n-k)!\,k!} $$

when $k = 2$.

½N(N−1) is also central to the discussion of any metric space: it is *the number of coefficients (or functions of position) required to define the metric properties of a space of N dimensions.*(Riemann, 2004, p 262f).

6.4 Number of Cross-Consistency Pairs

If the number of parameters in a morphological model is N and the number of parameter values for a parameter P_x is v_x, then the number of dyadic (pair-wise) relationships (Ct) between *all parameter values* (and thus the total number of cells in the cross-consistency matrix – CCM) is:

$$ \mathrm{Ct} = \sum_{i=1}^{n-1} \sum_{j=2}^{n} v_i \cdot v_j. $$

Table 6.1 The primary formal properties of morphological models (for $v = 4$)

a	*b*	*c*	*d*
N	½N(N−1)	$\sum_{i=1}^{n-1}\sum_{j=2}^{n} v_i \cdot v_j$	$\prod_{i=1}^{n} vi$
Number of parameters	Number of dyadic relationships between parameters blocks	Number of CCM cells	Number of simple configurations
2	1	16	16
3	3	48	64
4	6	96	256
5	10	160	1,024
6	15	240	4,096
7	21	336	16,348
8	28	448	65,536
9	36	576	262,144

The take-home message is this: that while the number of formal configurations in a morphological model increases "geometrically" (factorially) with each additional parameter, the number of cross-consistency pairs increases "only" in proportion to the quadratic polynomial f(x) = ½x(x-1). This is what makes it possible to employ Cross-Consistency Assessment (CCA) in order to reduce a relatively large problem space to a more manageable solution space, without having to examine every configuration in the problem space.

To sum up: we have four magnitudes which determine the primary formal properties of a morphological model:

N = number of parameters
½N(N−1) = number of parameter blocks
ΣΣ vv = number of pair-related cells in the CCM
Π v = total number of simple configurations in the model

In the case of v = 4 for each parameter, the relationship between these magnitudes as given in Table 6.1.

6.5 Three Ratios

Expressions *b*, *c* and *d* (Table 6.1) are formally determined by N and V_x, i.e. the number of parameters and the number of conditions under each of the parameters. There are three other quantities that are determined by the *empirical properties* of the Cross-Consistency Assessment (CCA) which, together with *b*, *c* and *d*, give rise to three ratios that can help us to formally "type" morphological models.

These three ratios are:

1. The *connectivity quotient* (**κ**): The ratio of the number of *constrained* parameters blocks (PBc) to the *total* number of parameter blocks ½N(N−1). This is analogous to how the dimensions of an abstract space are topologically connected.
2. The *consistency quotient* (**χ**): The ratio of the number of *mutually constrained parameter value pairs* in the Cross Consistency Matrix (CCM) to the *total number of parameter value pairs* (or *cells*) in the CCM.
3. The *solution space quotient* (**ζ**): The ratio of the number of simple configurations in the *solution space* to the number of simple configurations in the *total problem space*.

6.5.1 Connectivity Quotient (κ)

Connectedness in a morphological model concerns how the different *dimensional constructs* of the model (i.e. its parameters) "hang together". That is to say, how does each parameter relate to each of the other parameters?

There are two (principal) possibilities here: either two given parameters are (logically and empirically) *orthogonal*, or they contain *mutual* (logical and/or empirical) *constraints*. (A further distinction can be made on the basis of the difference between *logical* and *empirical* constraints.)

Orthogonal means "at right angles". This means that the value ranges of two orthogonal parameters are independent of each other, i.e. they do not interfere with or constrain one another. Since in morphological modelling we relate values by way of mutual consistency, then we say, that in a pair of orthogonal parameters P_a and P_b, any value of P_a is consistent with (can co-exist with) any value of P_b. Figure 6.3 shows an orthogonal parameter pair (block), in which the assessment key "-" means: "is consistent with…" or "can co-exist with …".

An orthogonal relationship between two parameters does not necessarily mean that there is no meaningful *content* associated with the value relations. It simply means that there are no mutual constraints between the parameters. However, if a parameter P_k is *orthogonal to all of the other parameters in a morphological model,*

		Pa			
		a1	a2	a3	a4
Pb	b1	-	-	-	-
	b2	-	-	-	-
	b3	-	-	-	-
	b4	-	-	-	-

Fig. 6.3 Orthogonal parameter block

Fig. 6.4 Empirically constrained parameter block

		Weight (Kg)			
		< 20	20-50	50-100	> 100
Age (Yrs)	< 5	-	-	?	X
	5-10	-	-	-	?
	10-20	?	-	-	-
	> 20	X	?	-	-

then its variability has *no effect on the rest of the model*. Such a parameter is – so to speak – exogenous to the model as such.

Parameter pairs are *mutually constrained* when *at least* one value pair in the parameter block is deemed inconsistent, impossible or unviable. For instance, if we pit a range of age intervals in a population against a range of body weight intervals, then obviously (for us *homo sapiens*), there are going to be some expected constraints between age values and weight values (as seen in Fig. 6.4). In this hypothetical example, for instance, "X" could mean *highly unlikely* and "?" *pretty extreme*. The diagonal area from bottom right to top left (containing "-") we could call the main sequence of the relationship. This type of pattern often turns up when scales are pitted against each other.

6.5.2 Connected vs. Unconnected Parameters

We say that two parameters are *connected* when they have mutual constraints between their value ranges. If a parameter block is not "connected", then it is "orthogonal". The connectedness between two parameters is not *directional*, but simply denotes that two parameters constrain or interfere with each other. The totality of the connections between parameters in a morphological model can be represented as an undirected *graph*.

The degree of connectedness of a morphological model (measured by the Connectivity Quotient $\boldsymbol{\kappa}$), is the ratio of the number of connected parameter blocks to the *total* number of parameter blocks in the model. However, in order to define a proper model – i.e. one that "hangs together as a whole" – then every parameter *must be connected to at least one other parameter*.

If the number of parameters in a morphological model is N, then the *minimum number* of connections for the model to hang together as a whole is $N-1$. The *maximum number* of possible connections between parameters is (as we know) ½N ($N-1$).

Thus an N-dimensional model is called *minimally connected* when it has $N-1$ connections and each parameter is *connected to one and only one other parameter*.

A model is called *completely connected* when it has ½N(N−1) connections, and every parameter is connected to every other parameter. While there is only one way to *completely connect* a model (i.e. every parameter is constrained by every other parameter), there are many ways to minimally connect a model.

To find out how many ways one can produce a minimally connected model, one needs to find *the number ways of picking k unordered outcomes from n possibilities.*

$$ {}_{n}C_{k} \equiv \frac{n!}{(n-k)!\,k!}. $$

For instance, if we take the four letters *abcd*, and want to see how many combinations of *three* letters we can make of this, then:

$$ {}_{4}C_{3} = \frac{4*3*2*1}{1*3*2*1} = 4 = (\text{abc}, \text{adb}, \text{acd}, \text{bcd}). $$

Note also, that when $k = 2$ – i.e. when we are making pair-wise comparisons – then ${}_{n}C_{k}$ always reduces to ½N(N−1).

Table 6.2 shows the quantitative relationships of connectivity in morphological models of N dimensions. The last column – the number of ways to produce minimally connected models – is only a curiosity. The morphologist is not concerned with this issue. The model will be connected on the basis of the logical, empirical and normative *content* of the model.

However, how parameters in a model are *actually connected* is of the utmost importance for the morphologist. The *Connectivity Quotient* **κ** is the ratio of the number of *constrained parameter blocks* (PBc) to the total number of parameter blocks ½N(N−1).

$$ \kappa = \frac{\text{PBc}}{\frac{1}{2}\text{N}(\text{N}-1)} $$

Table 6.2 Relationship of connectivity for an N-dimensional morphological model

Parameters N	Minimal connections N−1	Total connections ½N(N−1)	Number of ways to produce minimally connected models
2	1	1	1
3	2	3	3
4	3	6	20
5	4	10	210
6	5	15	3,003
7	6	21	54,264
8	7	28	1,184,040
9	8	36	30,260,340

Since the minimum number of Constrained Parameter Blocks (PB_c) required to define a proper model is $(N-1)$, then the possible range of PB_c is:

$$(N-1), \frac{1}{2}N(N-1).$$

Thus κ ranges from:

$$\frac{(N-1)}{\frac{1}{2}N(N-1)}, \ 1 = \frac{2}{N}, \ 1.$$

6.5.3 *The Consistency Quotient (χ)*

The *consistency quotient* is the ratio of the number of mutually constrained (i.e. inconsistent) cells (Cx) in the Cross-Consistency Matrix (CCM) to the total number of cells (Ct) in the CCM.

$$\chi = Cx/Ct,$$

where

$$Ct = \sum_{i=1}^{n-1} \sum_{j=2}^{n} v_i \cdot v_j.$$

The number of pair-wise mutually constrained cells (Cx) in a cross-consistency matrix is determined by the judgements made by the *subject specialist group* doing the morphological modelling. It is an "empirical" input, in the sense that it is not determined by any formal properties of the model. Rather, it is determined by the explicit or implicit *nature of the concepts* supplied in order to create the model. In order to determine Cx, one simply has to count them in the CCM.

6.5.4 *The Solution Quotient (ζ)*

The solution quotient ζ (*stigma*) is the ratio of the number of simple configurations making up the *solution space* ($\text{Config}_{\text{sol}}$) to the total number of (formal) simple configurations in the *problem space*.

$$\zeta = \text{Config}_{\text{sol}} \ [\textbf{divided by}] \ \prod_{i=1}^{n} v_i.$$

6.6 The Relationships Between the Three Ratios

Intuitively, we would expect there to be some sort of pattern between these relationships, since, clearly, **κ** and **χ** should influence the size of the solution space, and thus the magnitude of **ζ**. Ultimately, we want to see how these relationships express themselves in different types of morphological models with different empirical contents.

The only way to do this is to select and collate a number of actual morphological models developed in real settings for real problems. We have selected 16 representative models, chosen for their spread and breadth of application. For the moment, the *purpose* of these models, i.e. their content, is not of any relevance. We simply want to see the spread of their formal characteristics as concerns the three cited ratios.

The collation is shown in Table 6.3 and Graph 6.1.

The graph is based on the idea that the relationship that best represents the formal constraining properties represented by the cross-consistency assessments (CCA) is: **χ [divided by] κ**, which we shall call the *constraining quotient*. This ratio expresses – in an approximate way – the distribution of the constraints over the whole cross-consistency matrix. The hypothesis is, that for a given **χ**, the constraints on the model will *tend* to be greater if these are concentrated to only a few parameter blocks, rather than being spread out more evenly over the entire matrix. Therefore, there should be some correlation between the constraining quotient (**χ [divided by] κ**), and the solution quotient **ζ**, which represents the relative size of the solution space.

The graph shows a sequence of models where **ζ** is plotted against (**χ [divided by] κ**). It shows a clear inverse relationship between the two, but not a very strong one. Clearly, models cannot deviate too far towards the upper right hand of the graph, as this would represent a highly constrained CCM producing a relatively large solution quotient – an obvious contradiction. It is the same story in the opposite direction: weakly constrained CCM:s would not be expected to produce very small **ζ**.

Intuitively, the divergences from this linear trend fit my expectations. What is needed at this point is a careful comparison of "divergent" models (e.g. those circled in Graph 6.1) in order to see what it is that determines this divergence. Experience tells me that two factors should contribute to such divergences: the scaling properties (including non-ordinal value ranges) employed in the dimensions of different models; and differences in the proportions of OR-list and AND lists employed in different models (see Chap. 4).

Unfortunately, I am forced to leave this question for a future article or, if there are enough morphologists around to care (and if I am still "around"), to a future edition of this book.

Finally, the obvious question: does the practicing morphologist have to know everything in this chapter in order to competently carry out a morphological analysis? The answer is: No.

Table 6.3 The formal properties of 16 selected models

Project	Parammeters N	Total Parameter blocks N (N−1)/2	Number constrained parameter blocks	(a) Connectivity quot. Constrainced blocks (divided by) total blocks (%)	Total CCA cells (Ct)	Number constrained cells (Cx)	(b) Consistency quot. Constrainced cells (divided by) total cells (%)	Total problem space (# configs)	Primary solution space (# configs)	(c) Solution quot. Size solution space (divided by) Total prob. space (%)	(d) Constraint quot. b/a
Org. structure	7	21	16	76	648	74	11	186624	2634	1.41	0.1499
Chemical accidents	8	28	12	43	410	60	15	57600	2032	3.53	0.3415
Chemical terror	10	45	22	49	651	71	11	518400	29948	5.78	0.2231
Multi-hazard	6	15	12	80	1218	143	12	550368	22906	4.16	0.1468
Cult structure	7	21	13	62	593	53	9	105000	5644	5.38	0.1444
Accident structure	8	28	15	54	648	76	12	544320	23581	4.33	0.2189
Transport distruptions	6	15	5	33	1471	212	14	534600	3960	0.74	0.4324
Nuke transport	4	6	5	83	423	92	22	3024	181	5.99	0.2610
Fuel ransion	5	10	8	80	330	77	23	6480	151	2.33	0.2917
Municipal housing	9	36	20	56	677	97	14	345600	1928	0.56	0.2579
Food footprint	5	10	5	50	874	271	31	91728	1867	2.04	0.6201
Migration model	5	10	4	40	1075	176	16	30240	270	0.89	0.4093
Pub-private coop	8	28	20	71	835	140	17	1008000	12178	1.21	0.2347
Post office-infra	8	28	18	64	615	103	17	235200	15586	3.83	0.2605
Post office-demands	8	28	20	71	678	133	20	588000	22520	3.83	0.2746
OXA	8	28	7	25	660	91	14	259200	3840	1.48	0.5515

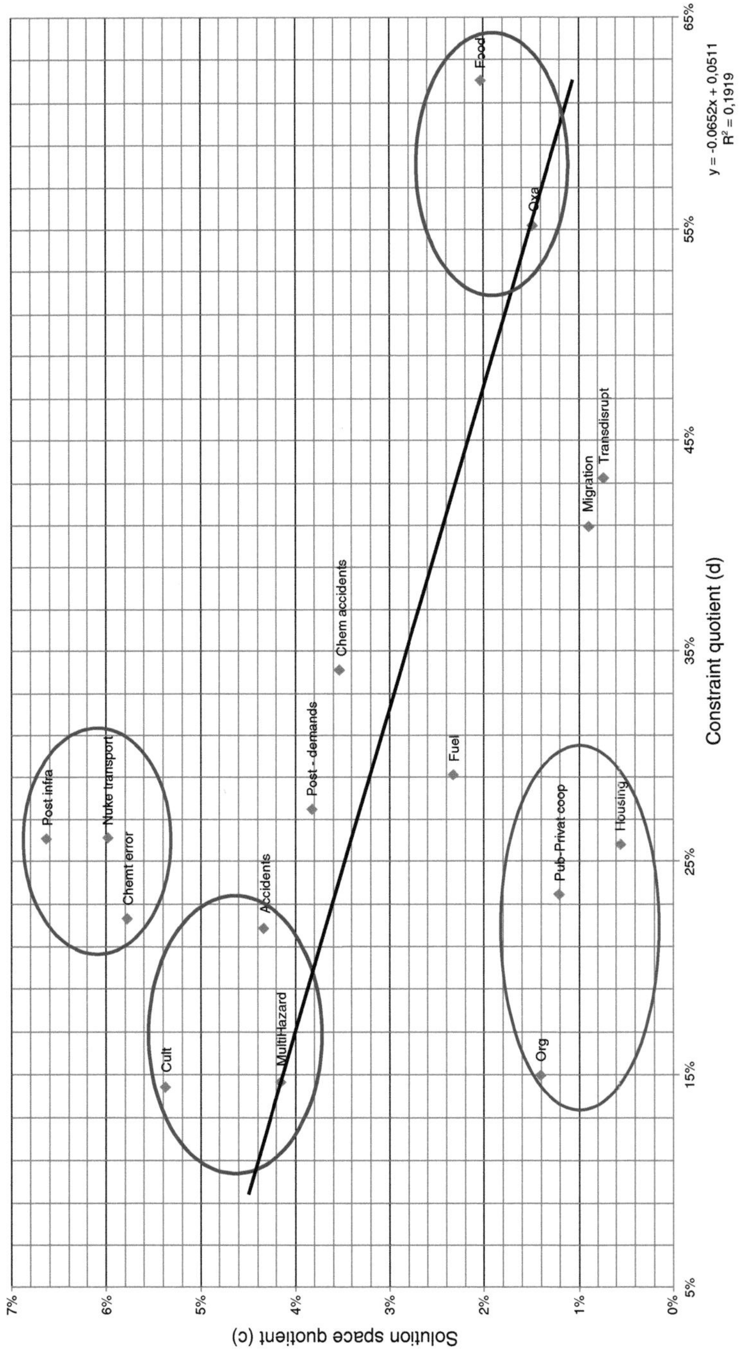

Graph 6.1 The constraint quotient X the solution space quotient

Of course, every morphologist will know that the problem space in a morphological model increases "geometrically" with each added parameter; that the number of Cross Consistency cells increases more "weakly", as a quadratic polynomial; and that there are obvious reasons for models being either hyper-coherent or hyper-constrained.

Above all, the competent morphologist – through years of experience and reflection – knows the difference between meaningful and trivial models; and also between what might be termed *well-behaved* models and *pathological* models. (Pathological models, which behave in bizarre and unexpected ways, much like "pathological functions" in mathematics, can be very meaningful indeed.)

Although the formal properties of morphological models are only derived and subordinate features of their content, for those of us who work in this field, these combinatorial issues are nonetheless thought-provoking.

I think that Fritz would have enjoyed this exploration.

Chapter 7
Facilitating GMA Workshops

7.1 Four Necessary Requirements

This is what I have learned during the past 15 years: There are four *necessary requirements* for carrying out GMA workshops for the purpose of creating (meaningful) morphological inference models concerning wicked problems:

1. Sound knowledge and experience of morphological modelling methods: i.e. theory, procedures, techniques, pitfalls etc.
2. Extensive knowledge and practical experience in small group facilitation and facilitation methods.
3. An appropriate working group of relevant Subject Matter Specialists (SMS)-cum-Stakeholders.
4. Dedicated, *flexible* GMA software.

Since I contend that all of these requirements are *necessary*, it would seem pointless to attempt to rank them. However, in my experience, the *least important* of these four necessary requirements is the *software*.

Given the most wonderful software in the world (presently MA/Carma™), a person lacking one or more of the first three requirements will risk making a complete disaster out of the GMA process. On the other hand, given the first three requirements, a competent morphologist/facilitator can pull off an acceptable GMA workshop – at least in the sense of *framing* a complex problem area – without dedicated software.

(What one cannot do without proper software is to produce interactive inference models that will allow the client/user to literally "play with their problem": i.e. look at it from different stakeholder perspectives, ask it "what-if" questions, change its initial conditions and link it to changing developments and environments.)

The *least appreciated* aspect of these requirements is the *facilitation of a SMS/Stakeholder group*. Certainly one can create morphological models "solo": This has been done for decades and is referred to as "attribute listing". It can be carried out using Excel sheets or any other spread sheet program. It is most usually done when one already knows (more or less) what a problem consists of

T. Ritchey, *Wicked Problems – Social Messes*, Risk, Governance and Society 17,
DOI 10.1007/978-3-642-19653-9_7, © Springer-Verlag Berlin Heidelberg 2011

"parametrically", and one wishes to explore or speculate on how these parameters are entangled. Do it. More power to it!

However, attempting to do a "solo" or "back-office" morphology, when dealing with really *wicked problems*, completely misses the point. GMA is a method for collective concept exploration, group creativity and the development of collective understanding of a complex problem area. Through dialogue and an exchange of ideas, it is intended both to bring forth tacit knowledge from SMS/stakeholder groups, and to foster the creation of new concepts and contexts.

There is plenty of evidence that facilitated group interaction consistently surpasses individual capacity in the area of concept exploration and creativity for "open ended" problems. (see e.g. Blinder & Morgan, 2005).

7.2 The Facilitation Thing

Michael Wilkinson (2004) calls it the *Fundamental Secret of Facilitation* (and then goes on to point out that it is not a secret at all; it is just not taken seriously enough). I think that is better called the *Fundamental Principle of Facilitation*. To paraphrase:

> *In dealing with complex societal problems, far more effective results will be achieved when these problems are framed, and solutions created and understood, by the people who are actually impacted by the problems – i.e. its various stakeholders.*

As we have seen, *wicked problems* are about stakeholder positions. And where stakeholder "buy-in" and a collective understanding of issues and positions are important, then *facilitation* becomes very important.

The art and science of facilitation began to emerge as a *discipline in its own right* only in the late 1960s. As such, it is still sometimes confused with the roles of *consultants* and *trainer-educators*. Check out the following definitions:

> "Facilitation is the process of enabling groups to work cooperatively and effectively. . . . . In particular . . . where people of diverse backgrounds, interests and capabilities work together." From *Information and Design*[1]
>
> [Facilitation is] ". . . the use of a *neutral* to help a group of people conduct productive discussions about complex or potentially controversial issues. The focus of the facilitator's role is to help people communicate effectively with each other." From the U.S. *Environmental Protection Agency*[2] (emphasis added).
>
> "A facilitator is someone who helps a group of people understand their common objectives and assists them to plan to achieve them *without taking a particular position in the discussion.*" From *Wikipedia*[3] (emphasis added).

This is not the place for a detailed exposition of the discipline of facilitation. There is a flora of books and articles on the subject, and plenty of material on the

[1] http://www.infodesign.com.au/ftp/Facilitation.pdf

[2] http://www.epa.gov/ne/enforcement/adr/glossary.html

[3] http://en.wikipedia.org/wiki/Facilitator

Internet. There is an International Association of Facilitators (IAF) as well as a European branch of that organisation, providing training material, discussion forums and other resources.

While there is a number of more or less standard procedures and "things to do" both prior to a facilitated meeting, during the meeting and after the meeting, there is also a number of important (and well-recognized) "dos" and "don'ts" for the facilitation process *in general*, which I have taken from the facilitation literature, especially inspired by Hogen (2002, 2003).

I learned these the hard way. Like most people my age, I started my facilitation career long before I even knew what *facilitation* was, and I started by making every mistake in the book. While these "dos" and "don'ts" may seem fairly clear cut, it is very easy to screw up when you start your facilitation career. (I exaggerate some of these "rules" a bit, in order to make a point. As in every discipline, you can sometimes break the rules, but only when you are experienced enough to know *when* and *how* to break them.)

7.3 Some Dos and Don'ts for New Facilitators

- As a facilitator, you are concerned with *form* and *process*; you *never argue content*! You are neither a SMS nor a stakeholder. If you are, then you have no business facilitating the group. (As far as content is concerned, you are like Manuel in *Faulty Towers*: *you know nothing*!).
- Be totally impartial. Never take sides or choose favourites. If you do, you're dead meat.
- Keep everything above-board. No hidden agendas.
- Engage; never manipulate.
- Acknowledge all inputs (ideas); it is not your job to assess them.
- Ask the group for help when you need it. Don't fake it! Admit mistakes and never become defensive.
- Get everyone to participate. (This will not be a problem is you choose the group correctly.)
- Encourage diversity; do not expect or demand ("first order") consensus. Promote "second order" consensus (see Glossary: Consensus).
- Get the participants away from addressing you, and get them talking to one another. (It is natural for group participants – at least initially – to address and talk content with *you*, as though you are the "group leader". It's a delicate matter, but you have to put a stop to it.)
- Don't be a *prima donna*. This isn't about you. Make yourself inconspicuous when the group is discussing matters. (I usually have a number of chairs placed out around the room, where I can sit down and temporarily disappear.)
- Encourage having fun – when appropriate. If you are not having fun, then you shouldn't be in this business (because you are certainly not going to get rich).

7.4 Facilitating WPs with GMA

Facilitating GMA workshops for the collective framing of WPs has some special features – some of which are quite nightmarish at the beginning of one's GMA-facilitation career:

As a rule, facilitators seldom know very much, or even anything, about the subject matter which they are facilitating. This is how it should be: "*thou shalt not be too knowledgeable* about the subject you are facilitating". However, in working with WPs another problem arises: although real, grounded knowledge is often elusive, there is plenty of personal opinion and emotion surrounding the problem complex. You are going to have to deal with this.

Furthermore, the SMS group you are facilitating is often not sure of, or agreed upon, *what the actual problem is that they are supposed to be exploring*. Certainly, each participant is knowledgeable about certain aspects of the problem complex (that is why you and the client have chosen them), but nobody knows what the total problem space looks like, since – in almost all the cases I have worked with – this problem space has never been properly *formulated* (morphed) before. This is the point of Rittel's first criterion for WP:s.

Also, the participants come from different areas of the (as yet *amorphous*) problem area, represent different aspects of the problem complex, and often represent opposing stakeholder positions. They literally come from different "tribes": they don't have the same backgrounds; they don't speak the same language; and they don't have the same priorities.

Since no one is sure about what the actual problem (ultimately) entails, it is nigh on impossible to tell the client, or the SMS group, how much time and effort (e.g. the number of group workshop days) it will take to form and explore the problem space, synthesise a solution space, and formulate alternatives. But the client almost always wants to know this in advance.

The whole enterprise can be laden with uncertainty, angst and denial. What client wants to hear that they are sitting in the middle of a colossal mess and don't know what to do about it? Who wants to hear that they have "problems" that have no (traditional or *tame*) "solutions"?

These challenges can be overcome if the GMA endeavour is *framed properly for the client at the outset*. For this purpose, I am simply going to list three sets of "guidelines" that I follow for organising and carrying out facilitated GMA workshops. These are:

- ***General guidelines concerning the GMA process*** *(presented to the prospective client).*
- ***Guidelines for selecting the SMS (Subject Matter Specialist) workshop participants*** *(presented to the client).*
- ***Facilitation guidelines and workshop ground rules*** *(a contract that I explicitly present to and discuss with the SMS workshop group at the beginning of the workshop).*

7.5 General Guidelines Concerning the GMA Process Presented to the Prospective Client

1. **The establishment of the "Principal client contact"**. This contact person is usually the "buyer" of the GMA workshops but, in any event, has a vested interest in the success of the workshops for competent decision support. S/he is the main contact person with whom to collaborate in the total modelling-cycle process.
2. **Give 1–2 h presentation of GMA as a scenario & strategy modelling technique**, preferably to group of people supporting the principle contact person as well as potential workshop participants.
3. **Meeting with the *Principal client contact*** in order to discuss the number of planned workshops days, dates, venue, "focus question(s)" and (crucially) group composition. The discussions must concern *at least* the following:

    - A preliminary, generic "focus question" for each modelling context is to be formulated of the form:
    - "What are the most important factors (parameters/variables) concerning ... [*the client's problem area*]... and how are these factors related to each other (how are they *entangled*)".
    - The ideal number of workshop participants is 6–7.
    - The venue should be a meeting room for at least 15–20 people, i.e. at least twice as many as the number of participants in the workshop. The meeting room must be properly furnished and prepared.
    - No "observers" are allowed to be present during the group GMA working sessions.
    - GMA workshops are carried out in sequences of 2-day sessions. If several workshops are to be carried out, these must have an agreed time-lap between them (days or weeks) in order that the process is allowed to mature.

4. **Group selection and composition is carried out collaboratively by the principal facilitator and the client** (see following Sect. 7.6).
5. **Distribution of focus question**: A week before the first workshop is to take place, the "focus question(s)" along with a suitable article on GMA, is sent out to the members of the SMS working group. They are encouraged to think about the focus question(s) concerning the most important factors (variables) in the problem area. They are also told that they need not otherwise prepare for the workshops.
6. **Beginning the first workshop day**: GMA as a method is presented in detail to the SMS group with examples/case studies close (but not too close) to the current problem area. The preliminary focus question is brought up and discussed, and the group is asked if they are satisfied with it, or if they want to adjust it.

7. **Analysis phase – Development of the initial *Morphological Field***: Using only a white board (no computer at first), the facilitator works with the group to identify the most important dimensions/parameters/variables in the problem complex. As each parameter comes up on the white board, one or two examples of its value range (conditions) are given, in order to help clarify the meaning of the parameter.

   This first phase of the MA process is the most important one, and often the most demanding one, since there can be uncertainty – or complete disparity – about what the most important parameters are, and how they are to be expressed. The process of "giving form/shaping" the initial morphological field is iterative and can take a full workshop day or more, depending on the size and nature of the problem complex. This initial field represents the total "problem space" and can contain hundreds of thousands of configurations, i.e. formal solutions. (This is the most demanding part of the GMA process for the *facilitator*. It can take years of experience to learn to be comfortable with this process.)
8. **Synthesis phase – Cross-Consistency Assessment (CCA)**: The next step in the analysis-synthesis process is to reduce the total set of (formally) possible configurations in a "problem space" to a smaller set of internally consistent configurations representing a "solution space". If the morphological field contains less than c. 100,000 possible configurations, this process can be carried out within the 2-day workshop process.

   If the morphological field contains considerably more than c. 100,000 configurations, the CCA process may need more time, or it can be done indirectly, by defining a number of configurations (scenarios, strategies, systems or structures) in the morphological field, and checking each of these for internal consistency.
9. **Examine the *structure* and *coherence* of the morphological model**. When a prototype morphological model is completed and compiled, it must be examined carefully to establish its nature and properties – how it coheres and behaves. There are six steps to this examination (the details of which are available to clients and GMA workshop partners):

   - Model coverage/model linkage
   - Model coherence
   - Parameter Activity Check (PAC)
   - Identification of multiple boundary values
   - Identification of driver and multi-driver structure
   - Time-line analysis

10. **Define ranges of scenarios, strategies or other configurations**. Any number of configurations representing scenarios, strategies, system structures or stakeholder positions can be generated and defined within the model and related to one another. The model allows the user to define configurations using initial inputs, desired outputs, and with clustered variations.

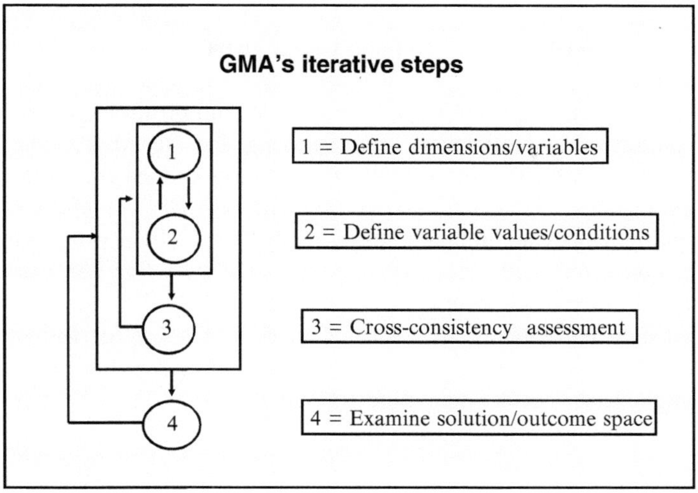

Fig. 7.1 GMA's iterative steps

The models belong to the client, who is supplied with dedicated software in order to run and maintain them. Six months "service" of the model is included in the workshop package.

The iterative process of developing a morphological model (see Fig. 7.1).

7.6 Guidelines for Selecting the SMS (Subject Matter Specialist) Workshop Participants

1. The ideal group size for morphological modelling is 6–7 subject matter specialist (SMS)/stakeholder participants. *No "observers" are allowed.* The group work is facilitated by 1–2 experienced GMA facilitators.
2. In choosing participants, identify the principal, general areas of competence that are needed for the mapping of the problem space. Avoid duplicate competencies and avoid competencies that are too specific or specialised (these can be brought in if and when needed).
3. All participants should be motivated. People who are "ordered" to participate, but do not really want to be there, will not contribute in an effective way.
4. Besides being experienced in their respective areas of competence, workshop participants should be intellectually curious and enjoy working with new methods, thinking out of the box and exploring new ideas *collaboratively*.
5 If possible, participants should come from different (relevant) areas within the organisation, or from different organisations, which *represent different aspects of the problem area and/or different stakeholder positions.*

6. Provided that the right competencies can be found, one should also strive for a good mix of gender and age. You don't want just a "bunch of old boys" sitting around the table.
7. Avoid people who represent high, strong leadership positions *if their presence* might inhibit other participants' free thinking and free discussion. No "big bosses". (In our defence sector work we regularly banned the participation of Generals.)
8. Avoid participants who think that they already "know all answers". GMA is not *initially* concerned with *"finding solutions"*, but to define *the total problem space of all possible solutions.*
9. All workshop participants are expected to show respect for and support the integrity of the working group as a whole. Facilitation is built on *confidence* and being able to "speak one's mind" openly.

7.7 Facilitation Guidelines and Workshop Ground Rules

(This is a *contract* that I explicitly present to and discuss with the SMS workshop group at the beginning of the workshop).

1. The facilitator is responsible for competently and correctly facilitating the *method and the process* and will not allow conditions to be imposed that would undermine the correct application of the method or the quality of the work.
2. The facilitator is not (and should not be) a subject matter specialist (SMS) or a stakeholder in the area relating to the workshop, or the results of the workshop.
3. The facilitator does not intrude upon the content or the *subject matter* of the workshop. The facilitator may ask questions of clarification concerning the concepts being used in the modelling process (sometime called "Socratic questions").
4. The subject matter specialists (SMS) *are responsible* for the content and subject matter relating to the workshop.
5. The facilitator will endeavour to bring all of the participants into the modelling process and discussions, and strive to bring all relevant issues "to the table".
6. There is no *voting* for allowing concepts to be discussed or brought into the model. Hidden agendas are discouraged and the censorship of ideas is disallowed. There are minority rights.
7. "Rules of Engagement" are discussed at the beginning of the workshop. A decision will be made on how to treat the process, the models and all other information generated by the workshop. Is this:

   (a) Open information?
   (b) Treated with the so-called Chatham house rule?
   (c) Not secret, but not for distribution?
   (d) Secret?

Chapter 8
GMA Case Studies

Eleven case studies are (briefly) presented here. They are:

1. Evaluating Preparedness for Chemical Accidents
2. Transport Disruption Scenario Laboratory
3. The Governance of Scientific and Technological Development
4. Anonymous Communication over the Internet
5. Nordic Energy Scenarios Framework
6. Electricity Grid Sabotage Scenarios
7. Multi-Hazard Disaster Reduction Strategies
8. Youth, Criminality and Social Exclusion in Sweden
9. Municipal Accident Strategies Model
10. Market Evaluation Template for a Govt. Authority
11. Modelling the Bioethics of Drug Redevelopment

Please Note. These case studies derive either from non-classified studies done for Swedish government agencies (and are therefore in the public domain) or have been released by the clients involved with certain conditions. In some cases, fields have been simplified or otherwise altered slightly at the client's request, in order to protect sensitive information. These will be noted, where relevant, for each of the case studies.

Also Note. The morphological modelling software we employ uses colour coding to differentiate between different aspects of the models, i.e. inputs, outputs, configuration comparisons, time-lines, etc. For reasons that are easy to guess, diagrams in this book are rendered in *grey scale*. Each diagram will thus be explained in these (grey scale) terms.

The first case study (*Evaluating Preparedness for Accidents Involving Hazardous Materials*) is presented in more detail in order to show how configuration comparisons are made and how to shift *input and output contexts*. The remaining

Some of these studies have been presented in earlier conference papers. For a list, see http://www.swemorph.com/references.html, under "Ritchey".

T. Ritchey, *Wicked Problems – Social Messes*, Risk, Governance and Society 17,
DOI 10.1007/978-3-642-19653-9_8, © Springer-Verlag Berlin Heidelberg 2011

case studies are presented in less detail, although the same procedures are applicable for all models.

8.1 Evaluating Preparedness for Chemical Accidents

Type of model: Duplex: Resource strategy *vs.* accident scenario.

For whom: Swedish National Rescue Services Agency.

When performed: 1999–2001.

Workshop participants: Two subject specialist teams (seven persons/team) of Fire Chiefs and Fire Engineers from five Municipal Rescue Services from different regions in Sweden.

Background and purpose: Swedish National Rescue Services Agency commissioned a study to develop a computer-based instrument for evaluating Swedish Rescue Services' preparedness for accidents involving hazardous materials (and also for terrorist actions involving the intentional release of chemical agents). The instrument should allow Municipal Rescue Services (MRS) to measure their preparedness resources and test their responses against different types of chemical accidents under different conditions, and to assign improvements to preparedness resources and see how these would improve response.

The model: The evaluation instrument for the Rescue Services is made up of two inter-linked fields: a general preparedness **Resource field** and a scenario specific **Response field**.

8.1.1 *Resource Field*

A rescue service's preparedness is described with the aid of a Resource field. This field is general, in that any and all rescue services – from part-time organisations in small municipalities to large metropolitan organisations – can be described within it. The resources are described by five parameters (the first five columns in the morphological field below).

8.1.2 *Response Field*

A Response field (the three rightmost columns in the matrix below) describes possible responses that a rescue service can make (depending on its resources) within a set of critical time periods *defined by a specific accident scenario*.

It is important to keep in mind that it is *preparedness* that we were assessing with this instrument, not what *might* be the actual outcome due to chance or outside influences.

All Response matrixes have the same general parameters:

- Responses associated with managing/controlling the release itself
- Responses concerning information dissemination to the public, and especially those in danger due to the release
- Responses associated with evacuation or rescuing people threatened or injured by the effects of the release

The exact formulation of the parameters, their order of priority and the "levels of response" expressed within them, are defined by way of specific accident scenarios. In the field show here, a scenario involving the release of a poisonous, pressurised gas, e.g. ammonia or chlorine, defines the response matrix. However, any number of scenarios can be developed, thus redefining the parameters and values of the response matrix, in order to test preparedness for different types of chemical accidents, or even other types of societal disruptions or disasters (Figs. 8.1–8.5).

On the basis of the result given in the response matrix for the particular scenario chosen, combined with the chosen geographical location of the accident, assess the possible consequences of such an accident and response. It is important to choose a relevant location and time of day etc. in order to test resources, and to provide a suitable challenge for the rescue organisation. Worst-case locations and times should be examined, as well as locations where accidents would be most expected to occur.

PLANNING/ PLANS	TRAINING AND EDUCATION	PERSONNEL AVAILABLE	EQUIPMENT AVAILABLE	COMMAND LEVEL	RESPONSE to chemical release	RESPONSE: Information to public	RESPONSE: Affected people
Full preparedness plan	Broad co-op. training	11 or more	Special equipment for specific case	Level 4	Reduce by least 80% within 15 min	Warn involved within 5 min	Help many within 30 min
Response plan for specific case	Training for specific case	8-10	Base equipment for specific case	Level 3	Reduce by least 80% within 30 min	Warn involved within 30 min	Help some individuals within 15 min
Standard routine for specific case	Base education + regular training	5-7	Less than base equipment for specific case	Level 2	Reduce by less than 50% within 15 min	No warning within 30 min	Help some individuals within 30 min
Standard routine for general case	Base education only	4 or less		Level 1	Reduce by less than 50% within 30 min		No help within 30 min
Only alert plan					No measures within 30 min		

Fig. 8.1 Chemical accident evaluation model with resource field (*white*) and response field (*grey*) for a specific chemical accident involving an ammonia release

		PLANNING/					TRAINING AN				PERSONNEL				EQUIPME			COMMAND				RESPONSE to					RESPONS		
		Full preparedness plan	Response plan for specific case	Standard routine for specific case	Standard routine for general case	Only alert plan	Broad co-op. training	Training for specific case	Base education + regular training	Base education only	11 or more	8-10	5-7	4 or less	Special equipment for specific case	Base equipment for specific case	Less than base equipment for specific case	Level 4	Level 3	Level 2	Level 1	Reduce by least 80% within 15 min	Reduce by least 80% within 30 min	Reduce by less than 50% within 15 min	Reduce by less than 50% within 30 min	No measures within 30 min	Warn involved within 5 min	Warn involved within 30 min	No warning within 30 min
TRAINING AND EDUCATION	Broad co-op.	-	-	-	X	X																							
	Training for	X	-	-	-	X																							
	Base education	X	X	-	-	-																							
	Base education	X	X	X	-	-																							
PERSONNEL AVAILABLE	11 or more	-	-	-	-	-	-	-	-	-																			
	8-10	-	-	-	-	-	-	-	-	-																			
	5-7	-	-	-	-	-	-	-	-	-																			
	4 or less	-	-	-	-	-	-	-	-	-																			
EQUIPMENT AVAILABLE	Special equipment	-	-	-	X	X	-	-	X	X	-	-	-	-															
	Base equipment for	-	-	-	-	-	-	-	-	-	-	-	-	-															
	Less than base	X	X	X	-	-	-	-	-	-	-	-	-	-															
COMMAND LEVEL	Level 4	-	-	-	-	-	-	-	-	-	-	X	X	X	-	-	-												
	Level 3	X	-	-	-	-	-	-	-	-	-	-	X	X	-	-	-												
	Level 2	X	X	-	-	-	-	-	-	-	X	-	-	X	-	-	-												
	Level 1	X	X	X	-	-	-	-	-	-	X	-	-	-	-	-	-												
RESPONSE to chemical release	Reduce by least	-	-	-	X	X	-	-	X	X	-	-	-	X	-	X	X	-	-	-	-								
	Reduce by least	-	-	-	X	X	-	-	X	X	-	-	-	X	-	X	X	-	-	-	-								
	Reduce by less	-	-	-	X	X	-	-	X	X	-	-	-	X	-	-	X	-	-	-	-								
	Reduce by less	-	-	-	-	X	-	-	-	-	-	-	-	X	-	-	-	-	-	-	-								
	No measures	-	-	-	-	-	-	-	-	-	-	-	-	-	-	-	-	-	-	-	-								
RESPONSE: Information to public	Warn involved	-	-	X	X	X	-	-	-	-	-	-	-	-	-	-	-	-	-	-	-								
	Warn involved	-	-	-	-	-	-	-	-	-	-	-	-	-	-	-	-	-	-	-	-								
	No warning	-	-	-	-	-	-	-	-	-	-	-	-	-	-	-	-	-	-	-	-								
RESPONSE: Affected people	Help many within	-	-	-	-	-	-	-	-	-	-	X	X	X	-	-	-	-	-	X	X								
	Help some	-	-	-	-	-	-	-	-	-	-	-	-	-	-	-	-	-	-	-	-								
	Help some	-	-	-	-	-	-	-	-	-	-	-	-	-	-	-	-	-	-	-	-								
	No help within 30	-	-	-	-	-	-	-	-	-	-	-	-	-	-	-	-	-	-	-	-								

Fig. 8.2 Cross-consistency matrix for accident scenario in Fig. 8.1

PLANNING/ PLANS	TRAINING AND EDUCATION	PERSONNEL AVAILABLE	EQUIPMENT AVAILABLE	COMMAND LEVEL	RESPONSE to chemical release	RESPONSE: Information to public	RESPONSE: Affected people
Full preparedness plan	Broad co-op. training	11 or more	Special equipment for specific case	Level 4	Reduce by least 80% within 15 min	Warn involved within 5 min	Help many within 30 min
Response plan for specific case	Training for specific case	8-10	Base equipment for specific case	Level 3	Reduce by least 80% within 30 min	Warn involved within 30 min	Help some individuals within 15 min
Standard routine for specific case	Base education + regular training	5-7	Less than base equipment for specific case	Level 2	Reduce by less than 50% within 15 min	No warning within 30 min	Help some individuals within 30 min
Standard routine for general case	Base education only	4 or less		Level 1	Reduce by less than 50% within 30 min		No help within 30 min
Only alert plan					No measures within 30 min		

Fig. 8.3 Model view with *resources* as input (*grey*) and *response* as output (*black*)

PLANNING/ PLANS	TRAINING AND EDUCATION	PERSONNEL AVAILABLE	EQUIPMENT AVAILABLE	COMMAND LEVEL	RESPONSE to chemical release	RESPONSE: Information to public	RESPONSE: Affected people
Full preparedness plan	Broad co-op. training	11 or more	Special equipment for specific case	Level 4	Reduce by least 80% within 15 min	Warn involved within 5 min	Help many within 30 min
Response plan for specific case	Training for specific case	8-10	Base equipment for specific case	Level 3	Reduce by least 80% within 30 min	Warn involved within 30 min	Help some individuals within 15 min
Standard routine for specific case	Base education + regular training	5-7	Less than base equipment for specific case	Level 2	Reduce by less than 50% within 15 min	No warning within 30 min	Help some individuals within 30 min
Standard routine for general case	Base education only	4 or less		Level 1	Reduce by less than 50% within 30 min		No help within 30 min
Only alert plan					No measures within 30 min		

Fig. 8.4 Evaluation model view showing initial configuration (*dark grey* and *black*) and augmented configuration (*light grey*)

PLANNING/ PLANS	TRAINING AND EDUCATION	PERSONNEL AVAILABLE	EQUIPMENT AVAILABLE	COMMAND LEVEL	RESPONSE to chemical release	RESPONSE: Information to public	RESPONSE: Affected people
Full preparedness plan	Broad co-op. training	11 or more	Special equipment for specific case	Level 4	Reduce by least 80% within 15 min	Warn involved within 5 min	Help many within 30 min
Response plan for specific case	Training for specific case	8-10	Base equipment for specific case	Level 3	Reduce by least 80% within 30 min	Warn involved within 30 min	Help some individuals within 15 min
Standard routine for specific case	Base education + regular training	5-7	Less than base equipment for specific case	Level 2	Reduce by less than 50% within 15 min	No warning within 30 min	Help some individuals within 30 min
Standard routine for general case	Base education only	4 or less		Level 1	Reduce by less than 50% within 30 min		No help within 30 min
Only alert plan					No measures within 30 min		

Fig. 8.5 Model view with *resources* as output (*black*) and *response* as input (*grey*)

8.2 Transport Disruption Scenario Laboratory

Type of model: Simplex: Mixed actor scenario model.

For whom: Swedish National Transport Administration.

When performed: 2004.

Workshop participants: Subject specialists from the four different Swedish transport authorities (road, rail, air and maritime), and from three public and private transport research institutes.

Background and purpose: The Swedish Transport Administration commissioned a study to develop a computer-based instrument for assessing how different *extraordinary events* in society can disrupt transportation, and how this disruption can propagate between and through the different transport functions/ sectors. The model was presented at a government sponsored conference attended by public and private carriers from all sectors in order to discuss common transport preparedness and hazard mitigation needs.

The model: The assessment instrument pitted a number of disruptive scenarios against vulnerable aspects of the four transport sectors plus joint (or intermodal) transport functions, and assessed the sectors' inter-related (second and third order) vulnerabilities to these disruptive scenarios. Four "single transport sector disruptions" were formulated as purely *analytic scenarios* to test how one transport sector shutdown would affect the remaining sectors. (The matrix of inter-related vulnerabilities remains restricted, and some of the parameters have been truncated.) (Figs. 8.6 and 8.7).

Scenario	Common transport functions	Disruptions in road traffic systems	Disruptions in rail traffic systems	Disruptions in maritime traffic systems	Disruptions in flight traffic systems
Intermittent electricity black-outs	Disruptions in inter-modal transport nodes	Disruptions in critical traffic routes	Regional stand-still	Disruptions/blocked national maritime routes	Disruptions in vital traffic control systems outside Sweden
Major snow storms during 2 weeks in greater Stockholm area	Disruptions in inter-modal booking and information services	Disrupted central traffic nodes	Major delays in entire country	Disruptions/blocked international maritime routes	International airports outside Sweden closed for Swedish traffic
Canada-like ice storm in southwest Sweden	Disruptions in inter-model follow-up systems	Lack of road capacity	Lack of engine and carriage capacity	Disruptions in critical ports and terminals	Disruptions in traffic control systems in route
Epidemic: Sweden in quarantine	Disruptions i customs and clearance functions	Lack of vehicle capacity	Lack of depot capacity	Disruptions in freight handling systems	Lack of aircraft and helicopter capacity
Major disruptions in tele-com in southern Sweden (antagonistic)	Normal function	Key personnel missing	Key personnel missing	Key personnel missing	Disruptions in vital airport functions
Analytic scenario 1 Major disruptions in road traffic		General shortage of personnel	Disruptions in IT/signal system	Disruptions in maritime traffic fuel supply	Key personnel missing
Analytic scenario 2 Major disruptions in rail traffic		Disruptions in road traffic fuel supply	Disruptions shunting and switch yards	Lack of tonnage	Disruptions in airline companies vital systems
Analytic scenario 3 Major disruptions in maritime traffic		Disruptions in information and C2 systems	Normal function	Disruptions in vital booking systems	Disruptions in aircraft fuel supplies
Analytic scenario 3 Major disruptions in flight traffic		Shortages of strategic product supporting road transport systems		Disruption/lack of towage, piloting servies	Lack of strategic products for airline sector
		Normal function		Disruption/lack of shipyard and repair services; spare part supplies	Disruptions in air rescue services C2 functions
				Normal function	Disruptions in the national aviation telenet (NATN)
					Normal function

Fig. 8.6 First order disruptions for the scenario "major snowstorms during 2 weeks in greater Stockholm area"

Scenario	Common transport functions	Disruptions in road traffic systems	Disruptions in rail traffic systems	Disruptions in maritime traffic systems	Disruptions in flight traffic systems
Intermittent electricity black-outs	Disruptions in inter-modal transport nodes	Disruptions in critical traffic routes	Regional stand-still	Disruptions/blocked national maritime routes	Disruptions in vital traffic control systems outside Sweden
Major snow storms during 2 weeks in greater Stockholm area	Disruptions in inter-model booking and information services	Disrupted central traffic nodes	Major delays in entire country	Disruptions/blocked international maritime routes	International airports outside Sweden closed for Swedish traffic
Canada-like ice storm in southwest Sweden	Disruptions in inter-modal follow-up-systems	Lack of road capacity	Lack of engine and carriage capacity	Disruptions in critical ports and terminals	Disruptions in traffic control systems in route
Epidemic: Sweden in quarantine	Disruptions i customs and clearance functions	Lack of vehicle capacity	Lack of depot capacity	Disruptions in freight handling systems	Lack of aircraft and helicopter capacity
Major disruptions in tele-com in southern Sweden (antagonistic)	Normal function	Ket personnel missing	Key personnel missing	Ket personnel missing	Disruptions in vital airport functions
Analytic scenario 1 Major disruptions in road traffic		General shortage of personnel	Disruptions in IT/signal system	Disruptions in maritime traffic fuel supply	Key personnel missing
Analytic scenario 2 Major disruptions in rail traffic		Disruptions in road traffic fuel supply	Disruptions shunting and switch yards	Lack of tonnage	Disruptions in airline companies vital systems
Analytic scenario 3 Major disruptions in maritime traffic		Disruptions in information and C2 systems	Normal function	Disruptions in vital booking systems	Disruptions in aircraft fuel supplies
Analytic scenario 4 Major disruptions in flight traffic		Shortages of strategic product supporting road transport systems		Disruption/lack of towage, piloting servies	Lack of strategic products for airline sector
		Normal function		Disruption/lack of shipyard and repair services; spare part supplies	Disruptions in air rescue services C2 functions
				Normal function	Disruptions in the national aviation telenet (NATN)
					Normal function

Fig. 8.7 First order disruptions for the analytic scenario "major disruptions in rail traffic" combined with disruption in one joint transport function

8.3 The Governance of Scientific and Technological Development

Type of model: Simplex: Problem structuring model.

For whom: Center for Science, Policy, and Outcomes (CSPO, Washington DC) and Columbia University (New York City).

When performed: 2002.

Workshop participants: Discussion groups formed during the symposium "Living with the Genie" at Columbia University, 2002.

Background and purpose: The Center for Science, Policy, and Outcomes (CSPO) commissioned an *in situ* study for the symposium "Living with the Genie" (Columbia University, 2002) in order to "model the conceptual problem space" dealt with at this symposium. The symposium treated the total issue of the governance of scientific and technological development: what it means and involves, its actors, variations, mechanics, ethics, measures and consequences.

The model: One of the prototype models "condensed" out of several discussion groups and plenum presentations. It is a simplex "problem space" of eight identified parameters. Because it was done *in situ* during a symposium, without a dedicated

Basic ruling principle for governance	Means or governance	Type of governance	Primary driving forces behind S&T development	Main trends in S&T	Principal funding	Who decides/ influences	Who benefits?
Human Rights	Spontaneous protests	Reactive	Higher values/ Spirit	Escalating change	Governmental	Politicians	Citizens
S&T a common good	Science shops	Proactive	To better our lot, Human needs	More complex language	NGO's	Companies	Monopolies
Monopoly	OTA (technology assessment)	Interactive	Social values	Larger groups	Companies	Public opinion	Companies
Limit harm	Patents		Curiosity	Monopolization	Venture capital	Courts	"The nation"
Intellectual property rights	Global funds/ programmes		Citizen security	Market		Organised intellectuals	Poor, developing nations
Efficiency	Change university curriculum		Short term profit	Individualisation		Media	The "global community"
Market	Education/ dialogue scientists-citizens		War			Local communities	
	Funding network to support diversity					Internal norms of a discipline	
	CBA					Funders	

Fig. 8.8 Problem space developed in situ at symposium "living with the genie"

focus group, dialog or feedback, no Cross-Consistency Assessment (CCA) has (yet) been performed on this field (Fig. 8.8).

8.4 Anonymous Communication Over the Internet

Type of model: Simplex: scenario model.

For whom: Swedish Ministries of Defence and Justice.

When performed: 2001.

Workshop participants: Subject specialists from areas of IT and communications technologies, an E-Commerce institute, the Faculty of Law at Stockholm University and the Justice department.

Background and purpose: The Swedish Ministry of Defence commissioned a study to explore future developments in Internet communication, especially the possibilities, incentives for and consequences of anonymous communication (AC). The study should include technical, economic, judicial, ethical and security aspect of anonymous communication.

The model: As is usual in such initial "problem structuring" processes, a number of models were developed in order to explore different aspects of the total problem area. One is presented here: a scenario model depicting different initiators and reasons for AC. Note that this study was performed almost 10 years ago. An awful lot has happed since then (Fig. 8.9)!

Scen	Who (Sender)	Who (Receiver)	Legal status of content	Anon. to whom	What is anonymised	Method of anonymisation	Who does anonymisation
Anon. love letter	Private person (individual)	Individual	Neutral or benign	Everybody	All identifiers of sender	Technical: with back door	Users themselves
The Leak	Collective (not legal person)	Closed group	Unethical	Everybody but receivers	Specific identity of sender	Technical: without back door	"Group operator"
The Freak	Association (Legal person)	"Open group"	Illegal	Receiver(s)	Relationships	Social	System operators
Terrorist planning	Business	Mass		Operator			Mix net, etc.
Blackmail	Public admin.						
Whistleblower							
The Gossip							
Business (anonymous PR)							

Fig. 8.9 Whistleblower scenario for anonymous communication study

8.5 Nordic Energy Scenarios

Type of model: Simplex: scenario framework model.

For whom: Energy Market Inspection Agency.

When performed: 2007.

Workshop participants: Subject specialists in the Nordic and European energy markets and energy technology from Sweden, Norway, Denmark and Finland.

Background and purpose: The Swedish Energy Market Inspection Agency commission a pilot study to develop a first modelling framework for developing a Nordic Energy Scenario Laboratory for a ten year futures period.

The model: This is a 13 parameter scenario framework for the initial problem structuring phase of a larger project. Such frameworks are usually broken down into smaller parameter groups and further analysed by different subject matter specialists. The initial framework was subsequently developed into an interactive scenario laboratory for developing common Nordic energy strategies (Fig. 8.10).

8.6 Electricity Grid Sabotage Scenarios

Type of model: Simplex: scenario model.

For whom: Swedish Energy Agency and Swedish Kraftnät (Power Net Utility).

When performed: 2000.

Legal framework in Europe	Regulation	Structure of generation companies	Overall economic developme nt in Europe	Energy price affect on Europe	Generation mix	Share of renewables	Nuclear power production	Ownership structure	Transmission system operator (TSO)	Unbundling	Number of PX in Europe	Principal PX- linking	Market power
Completely harmonised	Total European	Major pan-European	Very strong	Promotes sustained development	Totally homogenous	40%	Significant increase	Commercial: maximise profit	1 per country	Total unbundling	1 all European	Volume	Stronger market power
More harmonised than today	Regional	Regional	Slow steady growth	Status Quo	More homogenous than today	30%	Small increase	State owned	1 per region	More unbund. than today	1 per region	Price and volume mix	Status Quo
Status Quo	Member state	National champion	Sustained recession	Hinders development	Status Quo	20%	Status Quo	Non-profit org.	1 pan-European	Status Quo	1 per country	Price	Weaker market power
Less harmonised than today	None	Fragmented with countries			Less homogenous than today	10%	Decrease			Less unbund. than today	Each country has several		

Fig. 8.10 Scenario framework field for Nordic energy scenarios

Workshop participants: Subject matter specialists from the Swedish Energy Agency, the state Power Net Utility, Consultants and Security specialists.

Background and purpose: A commissioned study to explore the many aspect of possible sabotage of the Swedish national electrical grid, which in total consists of approximately 15,000 kilometres of 200 and 400 kV lines plus installations and interconnectors to neighbouring countries and IT systems.

The model: A nine parameter threat scenario model for antagonistic disruption of the Swedish electric power grid. Some 20 threat scenarios were identified and studied in order to define new Design Basis Threat levels for grid security (Fig. 8.11).

8.7 Multi-Hazard Disaster Reduction Strategies

Type of model: Simplex strategy model.

For whom: Earthquake Disaster Mitigation Research Center (EDM) – Kobe.

When performed: 2005.

Workshop participants: Seven subject matter specialists from the social, natural and engineering sciences who were participating in the World Conference on Disaster Reduction (Kobe) in January 2005.

Background and purpose: Multi-hazard *Disaster Risk Management* (DRM) is a complex problem area requiring expert knowledge and much practical experience

MOTIVE/ PURPOSE	GROUP SIZE	EXTENT OF NETWORK	LEVEL OF COMPETENCE (Truncated)	ETHICAL LIMITS (Intentional)	TYPES OF WEAPONS (Truncated)	TYPE OF OBJECT (Truncated)	SCOPE	Consequences
Personal revenge	A few individuals	Local	System - High Weapon - High	OK Nuclear	IT - system penetration	Conventional production plant	Local	1
Demonstrate superiority	Smaller groups	National	System - High Weapon - Low	OK RBC	Electro-mag.	Nuclear plant	Regional	2
Protest against social injustice	Large groups	International	System - Low Weapon - High	Indiscriminate killing	Hand tools	Operations center	National	3
Gain political advantage	Populations		System - Low Weapon - Low	Lethal force if required	Conventional weapons/explosives	Transformer station		4
Economic profit	Nation state			Injure people	PGM/ Guided missiles	Coupling station		5
Competition				Damage to property	Larger weapons systems	Cables		6
Disrupt society				Only stop infrastructure	Large vehicles/ Aircraft	Power lines		7
Gain control of territory					RBC-weapons			8
Group revenge					RBC-combination			9
Demonstrate power/politics					Nuclear weapons			

Fig. 8.11 Threat scenario model for disruption of Swedish electric power grid (some parameters truncated or generalised)

in a wide range of disciplines. It also requires a methodology which can collate and organize this knowledge through a participatory dialogue process. Towards this end, seven specialists from the social, natural and engineering sciences participated in a facilitated workshop to produce a prototype multi-hazard disaster reduction model which allows users to compare different hazards in terms of risk reduction strategies and adequate planning, preparedness and mitigation measures.

The model: This is a prototype or feasibility model intended to demonstrate the possible of identifying and comparing risk reduction strategies, and preparedness and mitigation measures, for different types of disasters. This would allow for the identification of synergies or disparities in disaster reduction methods as concerns different types of hazards, which may be concurrent. It also provides a common conceptual framework and terminology over a wide range of disaster reduction issues. The prototype was a condensed form of the original 11 parameters (Fig. 8.12):

1. Types of hazards
2. Principle risk reduction strategies
3. Root causes of vulnerability
4. Adequate knowledge required
5. Adequate planning measures
6. Adequate mitigation measures
7. Adequate preparedness measures
8. Legal/institutional frameworks needed

Hazard (Examples)	Risk reduction strategies	Unsafe physical conditions & practices which help propagate hazard	Adequate mitigation measures	Adequate preparedness measures	Adequate planning measures
Earthquake	Prevent the hazard from happening	Population density	Building standards for new construction	Warning systems	Risk analysis
Floods	Reduce severity of the hazard itself	Unsafe location	Building retrofit	Evacuation system	Information management & dissemination
Tornadoes	Reduce physical exposure to hazard	Lack of safe space	Land usage controls	Relevant education and training systems	Mitigation planning
Cyclones/hurricanes/ typhoon	Reduce consequences of hazard	Building vulnerability	Site level controls	Public awareness measures	Response planning
Fire	Reduce secondary hazards	Lack of adequate housing	Hazard control structures/works	Capacity enhancement	Recovery planning
Volcanoes	Risk transfer	Weak critical facilities and infrastructure	Infrastructure location & design	Contingency planning for critical facilities	Public involvement/ participation planning
Tsunamis		Weak institutions and legal framework	Content adjustments		Integration with development planning
Landslides		Lack of disaster planning	Relevant education & training		
Temperature extremes		Lack of provision for minorities and social equity	Natural environment protection		
Snowstorms/ ice storms		Lack of integration of planning and provision between systems levels	Development of livelihood security		
Urban drought		Lack of neighbourhood planning and provision, action	Application of low-cost and "appropriate technologies"		
Pandemic/epidemic		Prevalence of endemic diseases	Urban renovation		
Accidental RBC releases			Creation of incentives		
			Insurance and risk transfer		

Fig. 8.12 Prototype *multi-Hazard disaster mitigation model* comparing earthquakes and tsunamis for reduction of consequences and building vulnerability

9. Dynamic negative pressures
10. Dynamic positive pressures
11. Unsafe physical conditions and practices affected

8.8 Youth, Crime and Social Exclusion in Sweden

Type of model: Simplex: problem structuring model.

For whom: Swedish National Police Board (Rikspolisstyrelsen).

When performed: 2006–2007.

Workshop participants: Two groups of seven subject specialists from local police districts, Swedish universities and government research institutes, local municipal authorities, schools and social services. Plus a six-person group consisting of individuals with personal experience of social exclusion in Sweden.

Background and purpose: To develop a series of models that would explore the structure, variations, causes and effects of – and possible mitigation measures for – social exclusion in Sweden. Special emphasis was put on youth and criminality as a consequence of social exclusion.

The model: One of the models (of nine developed for the study) showing example groups (one selected) and possible exclusion profiles (Fig. 8.13).

8.9 Municipal Accident Strategies Model

Type of model: Duplex: Case – strategy alternatives model.

For whom: Swedish National Rescue Services Board.

When performed: 2002.

Workshop participants: Subject specialists from the National Rescue Services Board, local Rescue Services and local municipal authorities.

Background and purpose: The Swedish National Rescue Services Board commissioned a study to develop models to explore alternative *security policy directions* and *municipal risk management strategies* for accidents occurring in different contexts within a municipality.

The model: This (abridged) model pits accident victims and contexts against possible municipal security and risk management strategies, instruments and actors. This example shows one selected context and victim (Fig. 8.14).

8.10 Market Evaluation Template for a Government Authority

Type of model: Simplex: problem structuring/strategy model.

For whom: Swedish National Police Board (Rikspolisstyrelsen).

Who is excluded? (Examples from Swedish urban areas)	Primarily excluded from what? (Specific social contexts)	Why excluded? (From society's standpoint)	Why excluded? (From individual's standpoint)	What are their primary social networks? (Where ARE they included?)	Relationship to dominant social norms & codes	Primary motive for crime
Young boys astray (Age 12-16)	Consumer society's status aspects	Structural / institutional discrimination	"Wrong" social and cultural capital	Transnational ethnic, religious networks	Assimilated or Integrated	Ideology
Youth constellations (Age 16-18)	Economic security	Deliberate discrimination, intolerance	Low on knowledge and education	Area of residence (territorial)	Integrated within one's own stable and "accepted" culture	Basic provision
Young professional criminals (Age: 20+)	Swedish "Way of life"	Temporary, context- bound discrimination	Swedish language	A subculture	One's own culture in conflict with dominant society	Power
Girl gangs	Established conception of religion	Society creates unrealistic expectations	Suspicious of society's institutions	Peer group	Double exclusion (from one's own and from dominant culture)	Revenge
Heavy drug addicts	Swedish "language community"	Segregated dwelling	Mistaken conceptions and expectations about society	Nuclear family	Underground	Status Prestige
Criminal MC gangs	Dominate "appearance community"	Demands for Legal status	Conscious dissociation from society	Relatives, "clan"		Money as primary income
Young potential terrorist	Modern society's value-foundations	Demands on language and cultural competence		Working place		Solidarity
Right-wing extremists	The every-day world	Lack western education		No network		Frustration Feelings of impotence
Left-wing extremists	Not especially alienated	Low tolerance for "handicaps"				
		Don't know!				

Fig. 8.13 Social exclusion model showing one example of an exclusion profile

Where accident occurs	Victim	Municipality's principal security policy for different situations	Risk management strategy	Means (instruments) of control	What is to be influenced	Target groups	Actors who perform actions
Dwelling	Elderly	Equal security for all	Accident prevention	Information	Behaviour	Potential victim	Private person(s)
Traffic and traffic milieu	Other vulnerable groups	Equal security within a particular accident context	Injury prevention	Education	Technical safety levels	Indirectly concerned parties	Rescue service
Public buildings public places	Children	Support for vulnerable groups	Injury limitation	Counselling	Course of the accident	Organisational responsibility (e.g. owners)	Social authorities
Industry	Adults	Max. effectiveness for reducing general risk level	Spreading risk	Norms			Municipal administration
Water	Important property	Parry low probability events with major consequences	Recovery measures after accident	Subsidies			Police
Terrain	Critical infrastructure	Individual responsibility		Municipal budget			State
	Natural environment						Private firms
							Non-profit organisations

Fig. 8.14 Municipal accident strategy model showing case of elderly person in home environment

Client: type of organization	Relationship to client	Buying power & decision potential	Needed level of consultancy help	Needed type of product	Type and degree of control	Motive for working with client
Central government administration	Marriage	Big money Strong decision potential	Strategic Unspecified	Method (service)	Autonomous research	Glory
Central government authority	Strategic alliance	Money weak Strong decision potential	Strategic Specified	Method (product)	Result governed	Educational/ Instructive
Local government	Subcontractor	Big money Weak decision potential	Operative Unspecified	Decision support study	Manage general aim	Fun
Government owned company	"One shot"	Money weak Weak decision potential	Operative Specified	Expert subject knowledge	Management in detail	Bread & butter
"Big business" Multinationals	Competitor		"Nuts and bolts"	Person in situ		Future investment & PR
Trade association				Second opinion		Quick, easy income
Academic organization						Societal duty
Consultancy firm						
Small and middle sized business (SME)						

Fig. 8.15 Market evaluation model for government authority. Local government organisation selected as potential client

Type of drug redevelopment	Who develops/ sponsors	Ethical constraints due to ...	Barriers to initiation	Primary benefits
Failed NCE (small molecule)	Large pharma	Unpopular animal usage or other testing methods	Legislation favours large incumbent	Clinical (Efficacy)
Failed NBE (biologic)	Small pharma (speciality pharma)	Trial design ethics	Cost	Patient compliance/ acceptance
Old drug - new tricks (NCE)	Virtual pharma	Where to use	IP	Commercial benefits
Old drug - new tricks (NBE)	Government	Where to develop	Reimbursement limitations	PR/Reputation
Specials (small molecule)	NGO	Where to trial	Technological constraints	Advancement of science
Orphans (NCE)	Charitable orgs.	Cultural/religious mores	Ethical	Family/Carer benefits (holistic)
Orphans (NBE)			Territorial preferences	Medical regulatory incentives giving benefit
Life cycle management (NCE)			Dubious efficacy	
Life cycle management (NBE)			Commercial legislation	
			Drug regulation	
			Viable distribution channel	

Fig. 8.16 Example of a non-governmental organisation pursing limited development of a small molecular weight entity for a rare disease, but limited by costs

When performed: 1999.

Workshop participants: Subject specialists from government research organisation representing marketing, information and recruiting functions, research and product development programs and project leadership.

Background and purpose: To develop a client/product development model for possible organisational expansion and diversification of a traditionally non-market organisation, such as a government research organisation.

The model: One of three models for organisational development, this one treats of client types, relationships and product/service strategies. One example selected (Fig. 8.15).

8.11 Modelling the Bioethics of Drug Redevelopment

Type of model: Simplex: partial scenario model.

For whom: Strategy Foresight Partnership LLP, U.K. (SFP).
(Reported by Dr. Nasir Hussain, SFP).

When performed: February 2010.

Workshop participants: Subject matter specialists from the mid-size, speciality pharmaceutical sector and a sector specialist in law and ethics.

Background and Purpose: SFP commissioned this exercise as a prelude to a white paper to explore the question: Given the constraints of increased regulation, reduced innovation and stringent pricing controls, how does large pharma assure the ethical re-usage of existing drugs for alternative disease indications?

The Model: Initially, 14 parameters were generated concerning *inter alia* the developer, distribution channels, benefits, ethical constraints, commercial legislation and regulations, IP and licensing status, developmental costs, marketing approaches and patents. These were coalesced into five parameters as shown in Fig. 8.16. The configuration shows the example of an NGO "going it alone" to develop a product for a rare disease, but encumbered by costs.

Chapter 9
About Fritz Zwicky

Fritz Zwicky is not a household name in science today. He was not a super star of the likes of Einstein, Hubble or Oppenheimer. Yet his influence was significant – far more than his present-day lack of fame would suggest. He was one of the broadest and most inventive scientists of his time, and combined theoretical studies with eminently practical, humanitarian activities.

Zwicky was born in Varna, Bulgaria, in 1898, the son of a Swiss merchant. At the age of 6 he was sent to his father's ancestral district in Switzerland, Glarus, for schooling. Although expected to take up a career in commerce, Fritz' early bent for science apparently persuaded his father to allow him to study engineering instead.

In 1914 he moved to Zürich where he subsequently enrolled in the Swiss Federal Institute of Technology. There he switched to mathematics and experimental physics, wrote his examination essay for no one less than Herman Weyl, and in 1922 took his doctorate with a dissertation on ionic crystals. Three years later he moved to the California Institute of Technology in Pasadena to work with, among others, the great experimental physicist Robert Millikan.

From this point on, Zwicky more or less worked out of Pasadena, both as a faculty member of Caltech (1927–1968) and research director/consultant for Aerojet Engineering Corporation (1943–1961). He became Professor of Astrophysics at Caltech in 1942 and was a member of the staff of Mount Wilson and Palomar Observatories until his retirement in 1968.

Zwicky is primarily known for his work in astrophysics, and especially his comprehensive galaxy surveys. However, he thrived on investigating and theorizing about extreme phenomena and the boundary conditions. This led him both to develop a method for systematically investigating multi-dimensional problem complexes and to formulating a number of hypotheses which represented significant breakthroughs in astronomy.

Zwicky and Walter Baade were the driving forces behind acquiring and installing the first Schmidt telescope to be used in a mountain-top observatory – the famous 18-in. Palomar Schmidt – in 1935. Schmidt's revolutionary new telescope made it possible to photograph large areas of the sky quickly, with little

T. Ritchey, *Wicked Problems – Social Messes*, Risk, Governance and Society 17,
DOI 10.1007/978-3-642-19653-9_9, © Springer-Verlag Berlin Heidelberg 2011

distortion. Zwicky used it to make the first rapid survey of the heavens, mapping out hundreds of thousands of galaxies (now called the Zwicky Galaxy Database).

As a result of this, Zwicky discovered that galaxies tended to cluster, opening up a new chapter in the history of astronomy and cosmology. At the same time, he applied the so-called virial theorem[1] to the Coma cluster of galaxies and obtained evidence of unseen (or "missing") mass, thus starting off the debate on what is now called *dark matter* (The 18-in. Schmidt was later used by Gene and Carolyn Schumacher to discover the comet Schumacher-Levy 9, which smashed into Jupiter in July 1994).

Pursuing the idea that "bright novae" were of fundamental interest for determining the distance to far-off galaxies, he and Walter Baade coined the term supernova (Baade & Zwicky, 1934a). These, Zwicky proposed, marked the transition from ordinary stars to neutron stars – which he was the first to hypothesize – and were the origin of cosmic rays (Baade & Zwicky, 1934b). This was an amazing (and correct) triple hypothesis and was an important step in the still on-going project to determine the size and age of the (visible) universe. (Zwicky's neutron-star-hypothesis entered mainstream astronomy in the 1960s). In 1937 Zwicky proposed that galaxies could act as gravitational lenses.[2]

Besides numerous other contributions to astrophysics, Zwicky was active in the aerospace industry. Just after WWII, he was appointed head of the U.S. Air Force teams that went to Germany and Japan to evaluate wartime research on jet propulsion. He was subsequently awarded the Medal of Freedom by President Truman for his work. He was also director of research at Aerojet Engineering and was involved in the development of jet and rocket propulsion systems – for which he obtained a number of patents. He is credited by some as being the "father" of the modern jet engine.

He was also vice president of the International Academy of Astronautics and founder of the Society for Morphological Research, where he enthusiastically advanced "General Morphology" for some 30 years – between the 1940s until his death in 1974.

---

[1]"The virial theorem states that, for a stable, self-gravitating, spherical distribution of equal mass objects (stars, galaxies, etc), the total kinetic energy of the objects is equal to minus 1/2 times the total gravitational potential energy. In other words, the potential energy must equal the kinetic energy, within a factor of two." See http://www.astro.cornell.edu/academics/courses/astro2201/vt.html

[2]Einstein had originally calculated that stars could act as gravitational lenses, but that the focal point would be too short for the effect to be observed from earth. According to a news report in the Los Angeles Times in the 1930s (which I found in Zwicky's family scrapbook at the Fritz Zwicky Foundation in Glarus, Switzerland) the idea that *galaxies* could act as gravitational lenses was first conceived by a Los Angeles "dish washer" and amateur physicist. He allegedly contacted a physicist acquaintance at CalTech, who in turn contacted Einstein and Zwicky. According to the story, Einstein never answered, but Zwicky did the calculations, found it valid, and got the credit. I have no idea about the veracity of this story. However, it was also the Los Angeles Times, on January 19, 1934, that (ignorantly) lampooned Zwicky's neutron star hypothesis with a cartoon entitled "Be Scientific with Ol' Doc Dabble". Was there bad blood between Zwicky and the L.A. times? Was the "dish washer" story concocted?

Zwicky has been described as a notorious maverick in science, both brilliant and insufferable. There are scores of anecdotes about his deeds and manners, his "salty" attitude and abusive statements (no doubt embellished over time). He is "credited" with coining the term a *spherical bastard*, i.e. "a bastard no matter which way you look at him". There is also a rap poem written about him – "The Dark Matter Rap" by the astronomer David Weinberg (1996). It begins:

> My name is Fritz Zwicky,
> I can be kind of prickly,
> This song had better start
> by giving me priority.
> Whatever anybody says,
> I said in 1933.
> Observe the Coma cluster,
> the red-shifts of the galaxies
> imply some big velocities.
> They're moving so fast,
> there must be missing mass!
> DARK MATTER!

However, despite his ascribed abrasive nature, Zwicky was in fact a great humanist. He was engaged in a number of charitable activities, including years of work to help rebuild scientific libraries destroyed during the Second World War and participating in the Pestalozzi Foundation's program to establish war orphan villages. In 1949, he was awarded the Presidential Medal of Freedom by Truman for his work on rocket propulsion during World War II.

The two Principal works in English by Zwicky on General Morphological Analysis are:

Zwicky, F. (1969). *Discovery, invention, research – through the morphological approach.* Toronto: The Macmillan Company.

Zwicky, F., & Wilson A. (Eds.) (1967). *New methods of thought and procedure: Contributions to the symposium on methodologies.* Berlin: Springer.

There are also two short biographical sketches in English:

Remembering Zwicky by Greenstein, J. (1964). *Engineering and Science, 37*, 15–19. (Available at: http://calteches.library.caltech.edu/354/2/zwicky.pdf)

Idea Man by Stephan M. Maurer, at Beamline: http://www.slac.stanford.edu/pubs/beamline/31/1/31-1-maurer.pdf

Chapter 10
Glossary of Terms

Please note: Most of these terms are not specific to General Morphological Analysis. Many of them have been borrowed from other areas of study and employed for the purpose of explaining and understanding the morphological modelling of complex societal problems. Their "meanings" may have shifted in emphasis. They are employed as "aids in understanding", not as dogma.

Audit trail

A sequence of records, each of which contains evidence or other forms of knowledge pertaining to and resulting from the execution of a process or system function. In the *morphological modelling* process it is the definitions of the *parameters* and *parameter values*, and the recording of the reasoning behind each of the *Cross-consistency assessments*.

Cross Consistency Assessment – CCA

Pertains to the process by which the *parameter values* (or parameter *conditions*) in the *morphological field* are compared with one another, pair-wise, in the manner of a cross-impact matrix. As each pair of conditions is examined, a judgement is made as to whether – or to what extent – the pair can coexist, i.e. represent a consistent relationship. This process reduces the total *problem space* to a smaller (internally consistent) *solution space*.

CARMA

***Computer Aided Resource for Morphological Analysis*:** A proprietary software system used for carrying out morphological modelling and creating non-quantified inference models.

T. Ritchey, *Wicked Problems – Social Messes*, Risk, Governance and Society 17,
DOI 10.1007/978-3-642-19653-9_10, © Springer-Verlag Berlin Heidelberg 2011

Cognitive Feedback

We use this term here to denote *causative action* on society (and the physical world) *as a result* of human cognitive processes and conscious self-reflection. CFB is essentially the opposite of *epiphenomenalism* – the doctrine that "mind" (or human consciousness) is only a by-product and has no influence or causal effect on the physical world. (Note: In psychology – especially in the area of decision making – *Cognitive Feedback* denotes a process that gives decision makers information about their own, and others, cognitive processes).

Complexity (in self-referential systems)

The term *complexity* is used here to denote a self-referencing system – e.g. a person, a social organisation, a community or society. This involves a level or domain of causality that goes beyond the notions of linear (mechanical) cause-effect. Such complex (self-referential) systems are *causal*, but not *determinant.* See Hofstadter (1979), Luhmann (1995) and Rittel (1972). A purely formal-mathematical model of self-reference is exemplified in Gödel's Incompleteness Theorems.

Complex Adaptive System (CAS)

In social science, a dynamic (dispersed and decentralized) network of agents (e.g. individuals, organisations, institutions, nations) acting concurrently and reacting to each other. Coherent behaviour in the system can arise from competition and cooperation among the agents themselves. However, order is emergent as opposed to predetermined the system's development is non-linear and irreversible and the system's future is generally unpredictable (see e.g. Holland 1994).

Coherence

Degree of interconnection and consistency between parts – in this case between the *parameter values* in a morphological field.

Conditions

(see also "Parameter Values"): The different states or values a parameter can take the parameter's value range (see Fig. 10.1).

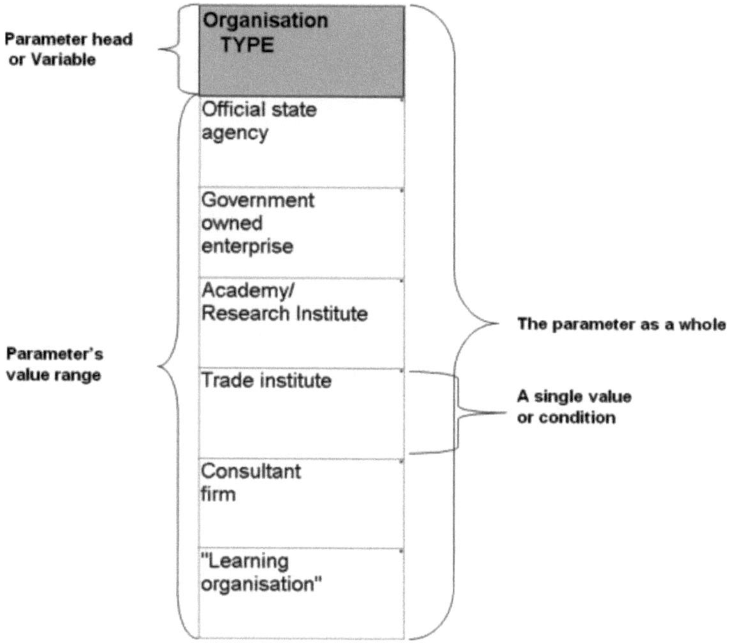

Fig. 10.1 Parameter terms

Consensus

Consensus usually means "general agreement or concord" within a group. Facilitators usually differentiate between *first-order* and *second-order* consensus. The normal, *first-order* form is that of gaining a common standpoint or agreeing upon a common solution. So-called *second order consensus* is when stakeholders in a group learn to accept each other's specific stakeholder positions – on the basis of understanding the *reasons* for these positions.

Consistency

Degree of compatibility between statements or conditions; in this case between the *conditions* of different *parameters* in a *morphological field.*

Configuration

At least one parameter value or condition displayed from each of the parameters in a morphological model (see Fig. 10.2).

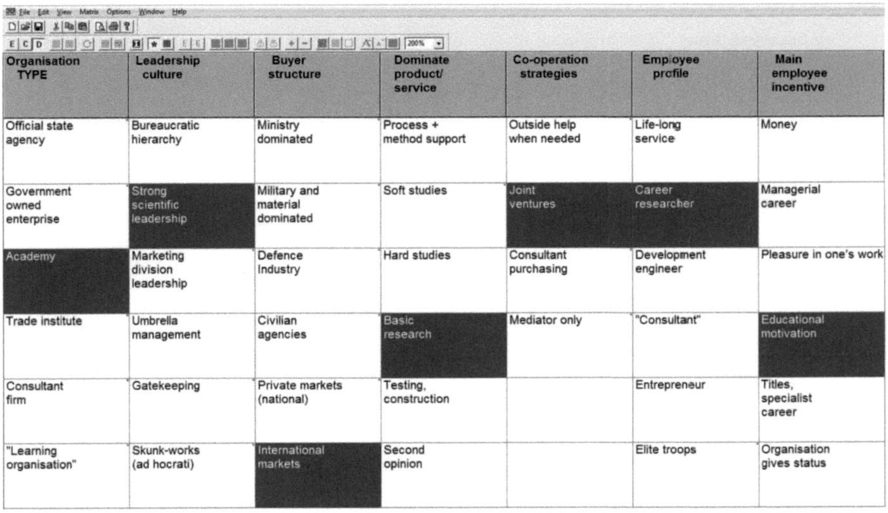

Organisation TYPE	Leadership culture	Buyer structure	Dominate product/ service	Co-operation strategies	Employee profile	Main employee incentive
Official state agency	Bureaucratic hierarchy	Ministry dominated	Process + method support	Outside help when needed	Life-long service	Money
Government owned enterprise	Strong scientific leadership	Military and material dominated	Soft studies	Joint ventures	Career researcher	Managerial career
Academy	Marketing division leadership	Defence Industry	Hard studies	Consultant purchasing	Development engineer	Pleasure in one's work
Trade institute	Umbrella management	Civilian agencies	Basic research	Mediator only	"Consultant"	Educational motivation
Consultant firm	Gatekeeping	Private markets (national)	Testing, construction		Entrepreneur	Titles, specialist career
"Learning organisation"	Skunk-works (ad hocrati)	International markets	Second opinion		Elite troops	Organisation gives status

Fig. 10.2 Morphological model with single configuration displayed

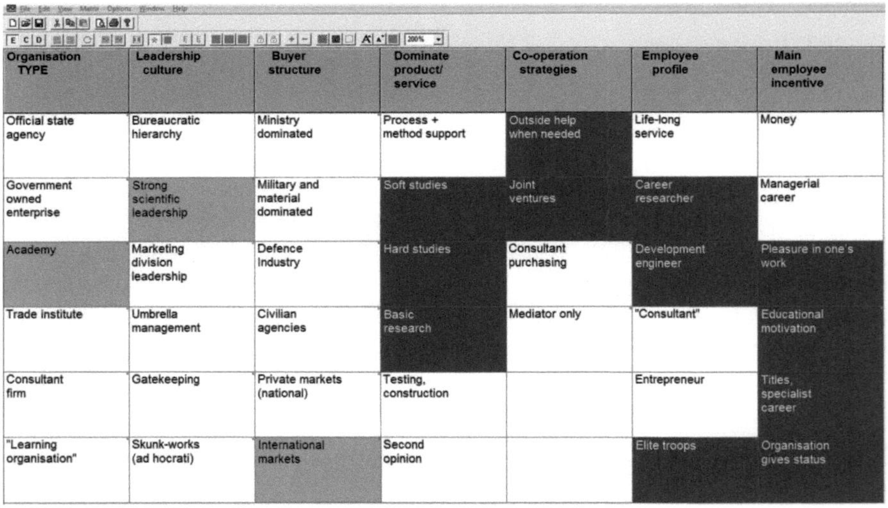

Organisation TYPE	Leadership culture	Buyer structure	Dominate product/ service	Co-operation strategies	Employee profile	Main employee incentive
Official state agency	Bureaucratic hierarchy	Ministry dominated	Process + method support	Outside help when needed	Life-long service	Money
Government owned enterprise	Strong scientific leadership	Military and material dominated	Soft studies	Joint ventures	Career researcher	Managerial career
Academy	Marketing division leadership	Defence Industry	Hard studies	Consultant purchasing	Development engineer	Pleasure in one's work
Trade institute	Umbrella management	Civilian agencies	Basic research	Mediator only	"Consultant"	Educational motivation
Consultant firm	Gatekeeping	Private markets (national)	Testing, construction		Entrepreneur	Titles, specialist career
"Learning organisation"	Skunk-works (ad hocrati)	International markets	Second opinion		Elite troops	Organisation gives status

Fig. 10.3 Morphological model with multi-driver input selected (*grey*) and clustered configuration displayed (*black*)

Configuration cluster

A Collection of configurations, all of which are consistent with a given (selected) condition (see Fig. 10.3).

Contextual Environment

Those processes and conditions in the outside world, which can influence our organisation, but which we cannot influence significantly ("External world factors").

Decision Support

Support for management decision making under uncertainty.

Decision Support System (DSS)

Software and modelling methods used to aid management decision making under uncertainty.

Dimension

A coordinate in any conceptual space, whereby a quantity or quality can be varied along a continuum or a discrete number of states. In conjunction with other such quantities or qualities, it serves to define the variables or degrees of freedom that determine a system's state or behaviour (see "Parameter").

Driver

A parameter that is of central importance to a process or model, and which tends to "drive" other parameters. A factor that influences many other factors, but is itself less influenced.

Empirical Inconsistency

A practical (empirical) incompatibility or discrepancy between two or more conditions or statements about the observed world (comp. Logical Inconsistency).

Field Coverage

Pertains to how much of the Morphological Field is covered by the solution space. One endeavours to have "full field coverage".

General Morphological Analysis (GMA)

A generalised form of Morphological Analysis which is not associated with any specific discipline. Developed by Professor Fritz Zwicky of the California Institute of Technology (CalTech) in the late 1940s. "I have proposed to generalize and systematize the concept of morphological research and include not only the study of the shapes of geometrical, geological, biological, and generally material structures, but also to study the more abstract structural interrelations among phenomena, concepts, and ideas, whatever their character might be." (Fritz Zwicky: Discovery, Invention, Research through the Morphological Approach, p. 34).

General Morphology

Another designation for General Morphological Analysis.

Hyper-Coherent

When the degree of compatibility or internal consistency between parameters in a morphological model is very high, and many possible solutions or outcomes are obtained (opposite of hyper-constrained).

Hyper-Constrained

When the degree of compatibility or internal consistency between parameters in a morphological model is very low, and very few possible solutions or outcomes are obtained (opposite of hyper- coherent).

Inconsistency

When two statements or conditions are logically or empirically incompatible or contradictory.

Influence diagram

In general, a qualitative model of a system, which depicts influence relationships between different elements or aspects of the system, shows the direction of such influences and (usually, but not always) allows for feedback loops or circular causality. In some cases, influences can be given relative strengths, and flows between nodes can be mapped. In other cases, the diagrams are only pictorial representations of complex nets of interaction.

Linkage (Linkage structure)

Concerns how parameters in a Morphological Field are linked, i.e. which parameters constrain each other, and which do not.

Logical Inconsistency (Analytic Contradiction)

A logical incompatibility or contradiction between two or more statements. A "contradiction in terms" (comp. Empirical Inconsistency).

Mess

(See: "Social Mess").

Model

A simplified, schematic representation of a system or phenomenon that accounts for its known or inferred properties and may be used for further study of its characteristics. Scientific models usually delineate the system's or phenomenon's variables and relate such variables to one another. In quantitative modelling these relations can be functional (mathematical) or statistical. In non-quantified modelling the relations can be logical (e.g. consistency) or pertain merely to influence (influence diagrams).

Morphology

The study of *form* or *structure* as such.

Morphological Analysis

The study of form or structure by identifying the multiple dimensions comprising any system, e.g. an organism, an organisation, a conceptual system or any entity taken as a whole. Employed in e.g. Zoology, Botany, Geology and Linguistics (See "General Morphological Analysis").

Morphological Field

The field of constructed dimensions or parameters which is the basis for a morphological model.

Morphological Model

A morphological field with its parameters assessed and linked through a Cross-Consistency Assessment (CCA).

Multi-Driver Inputs

Using multiple drivers as input to a morphological model in order to determine numerous possible outputs or results (see "Driver").

Normative Inconsistency (Normative Constraint)

An incompatibility or discrepancy between two or more conditions based on social norms, ethics and standards.

Parameter

One of a set of measurable factors that defines a system and determines its behaviour, and which can be varied in an experiment.

Parameter space

A set of mutually linked parameters making up the Morphological Field.

Parameter Values

(see also "Conditions"): The different states or values a parameter can take on the parameter's value range (see Fig. 10.1).

"Problem"

Use by Ackoff (1974) to denote a well-defined issue where the parameters and the mutual relationships between parameters are known, but where there is no single, unequivocal solution. Instead there are many possible solutions "depending on" how different aspects of the "problem" relate to each other.

Problem space

The totality of the possible configurations obtained in a Morphological Field (see "Solution space").

Problem Structuring Methods

A family of methods that apply modelling approaches to address messy or wicked problems faced by managers of organizations. These methods seek to alleviate or improve situations characterised by uncertainty, conflict and complexity.

"Puzzle"

Use by Ackoff (1974) to denote a well defined issue where the parameters and the mutual relationships between parameters are known, and where there is a single, unambiguous solution.

Risk

A form of uncertainty that has a well-grounded (quantitative) probability. Risk = (probability of something happening) × (consequences if it does happen).

Risk Analysis

Generally, the science of risks, their probability and evaluation. In business, it is a technique to identify and assess factors that may jeopardize the success of a project or achieving a goal. The technique also helps to define preventive measures to reduce the probability of these factors from occurring and to identify appropriate countermeasures.

Risk Mitigation

Long-term measures for reducing or eliminating risk.

Scenario

An outline or model concerning a hypothetical sequence of events (development over time) or a set of circumstances (futures projection). A scenario is usually related to a given time period (usually, but not always, the future); treats a number of defined variables (e.g. economic conditions, demographic conditions, etc.); and allows the variables to be related or linked to each other by way of different types of relationships (e.g. causal relations, probabilities, internal consistency).

Scenario Laboratory

A morphological inference model developed to map and interrelate scenario variables. It allows these variables to be related or linked to each other by way of internal consistency.

Social Mess (see also "Wicked Problems")

Used by Ackoff (1974) to denote a social or organisational planning problem that is vaguely defined, ambiguous and reactive (i.e. it reacts when you try to do something with it). These are primarily long-term social and organisational planning problems with many and varied stakeholders. (See: "Complex Adaptive System (CAS)").

Solution Space

The subset of all of the configurations in a morphological model which fulfil the requirement of being internally consistent, and thus being a possible solution.

Strategy Space

The internal world of an organisation, comprising those factors which the organisation can influence and mould into a strategy for coping with the contextual environment.

Transactional Environment

Factors which are external to an organisation as such, but which the organisation may be able to influence (e.g. through information campaigns, legal actions, lobbying, etc.).

Uncertainty (Genuine Uncertainty)

While risk is a form of uncertainty that has a well-grounded (quantitative) probability, *genuine uncertainty* cannot be ascribed a probability. In many cases, genuine uncertainty cannot be reduced to any significant extent and must be dealt with by, for instance, strategic flexibility (concerning the Contextual Environment) or proactive measures (concerning the Transactional Environment).

Values

See "Parameter values" or "Conditions".

Variable

A quantity or quality capable of assuming a set of values or conditions.

Wicked Problems: (See also "Social Mess")

Used by Rittel and Webber (1973) to denote a complex, continually developing and mutating social, organisational and policy planning problems.

References and Further Reading

Ackoff, R. L. (1974). *Redesigning the future: A systems approach to societal problems*. New York: Wiley.

Ackoff, R. L. (1979a). The future of operational research is past. *Journal of the Operational Research Society, 30*, 93–104.

Ackoff, R. L. (1979b). Resurrecting the future of operational research. *Journal of the Operational Research Society, 30*, 189–199.

Ayres, R. U. (1969). *Morphological analysis, technological forecasting and long-range planning* (pp. 72–93). New York: McGraw-Hill.

Baade, W., & Zwicky, F. (1934a). On super-novae. *Proceedings of the National Academy of Science, 20*, 254.

Baade, W., & Zwicky, F. (1934b). Supernovae and cosmic rays. *Physical Review, 45*, 138.

Bailey, K. (1994). *Typologies and taxonomies – an introduction to classification techniques, Sage University Papers*. Thousand Oaks: Sage Publications.

Bateson, W. (1896). *Materials for the study of variation.* Baltimore: Johns Hopkins University Press.

Blinder, A., & Morgan, J. (2005). Are two heads better than one? Monetary policy by committee. *Journal of Money, Credit and Banking, 37*(5), 789–812.

Bridgewater, A. V. (1969). Morphological methods – principles and practice. In R. V. Arnfield (Ed.), *Conference on technological forecasting, university of Strathclyde* (pp. 241–252). Edinburgh: University Press.

Britton, N. (2005). Developing a framework for identifying sound practices in megacities. 1st International Conference on Urban Disaster Research, 18–20 January 2005, Kobe, Japan.

Churchman, C. W. (1968). *The systems approach.* New York: Delta.

Conklin, J. (2001). Wicked problems and social complexity. *CogNexus Institute.* Available online at: http://cognexus.org/wpf/wickedproblems.pdf (accessed 2010.12.01)

Coyle, R. G., & McGlone, G. R. (1995). Projection scenarios for south-east Asia and the southwest Pacific. *Futures, 27*(1), 65–79.

Coyle, R. G., & Yong, Y. C. (1996). A scenario projection for the South China sea. *Futures, 28*(3), 269–283.

De Waal, A., & Ritchey, T. (2007). Combining morphological analysis and Bayesian networks for strategic decision support. *ORiON, 23*(2), 105–121.

Derman, E. (2004). *My life as a quant – reflections of physics and finance*. New York: Wiley.

Dilthey, W. (1989 [1883]). Introduction to the human sciences. Princeton: Princeton University Press.

Doty, D. H., & Glick, W. (1994). Typologies as a unique form of theory building. *Academy of Management Review, 19*(2), 230–251.

T. Ritchey, *Wicked Problems – Social Messes*, Risk, Governance and Society 17, DOI 10.1007/978-3-642-19653-9, © Springer-Verlag Berlin Heidelberg 2011

Fernandez, A., Britton, N., & Ritchey, T. (2006). Application of a prototype morphological model for earthquake disaster risk management. *Presented at the 2nd Asian Conference on Earthquake Engineering 2006, March 10–11, 2006*. Manila.

Gärdenfors, P. (2004) Conceptual Spaces: The Geometry of Thought. The MIT Press.

Greenstein, J. (1974). Remembering Zwicky. *Engineering and Science, 37*, 15–19.

Hanisch, C. (2000). Is extended producer responsibility effective? *Environmental Science & Technology, 34*, 170A–175A.

Hofstadter, D. G. (1979). Gödel, Escher, Bach: An Eternal Golden Braid. Basic Books.

Hogen, C. (2002). *Understanding facilitation: Theory and principles*. London: Kogan Page.

Hogen, C. (2003). *Practical facilitation: A toolkit of techniques*. London: Kogan Page.

Holland, J. (1994). *Complexity: The emerging science at the edge of order and chaos*. Harmondsworth: Penguin.

Horn, R. (2001). Knowledge Mapping for Complex Social Messes. A presentation to the "Foundations in the Knowledge Economy" at the David and Lucile Packard Foundation.

Kirby, M. W. (2003). The intellectual journey of Russell Ackoff: From OR apostle to OR apostate. *The Journal of the Operational Research Society, 54*, 1127–1140.

Knight, F. H. (1921). *Risk, uncertainty and profit*. Chicago: Houghton Mifflin Company. Available online at: http://www.econlib.org/library/Knight/knRUP1.html, § I.I.26. (accessed 2009.11.14)

Luhmann, Niklas. (1995) *Social Systems*. Stanford University Press.

McGhee, G. R. (1999). *Theoretical morphology: The concept and its applications*. New York: Columbia University Press.

Müller-Merbach, H. (1976). The use of morphological techniques for OR-approaches to problems. *Operations Research, 75*, 27–139.

Norris, K. W. (1963). The morphological approach to engineering design. In J. C. Jones & D. G. Thornley (Eds.), *Conference on design methods* (pp. 115–140). Elmsford: Pergamon Press, Inc.

Pike, K. (1954). *Language in relation to a unified theory of the structure of human behaviour*. Glendale: Summer Institute of Linguistics.

Rhyne, R. (1971). *Projecting whole-body future patterns – the field anomaly relaxation (FAR) method*. Menlo Park: Educational Policy Research Center of Stanford Research Institute.

Rhyne, R. (1981). Whole-pattern futures projection, using field anomaly relaxation. *Technological Forecasting and Social Change, 19*, 331–360.

Rhyne, R. (1995a). *Evaluating alternative Indonesian sea-sovereignty systems*. Informs: Institute for Operations Research and the Management Sciences.

Rhyne, R. (1995b). Field anomaly relaxation - the arts of usage. *Futures, 27*(6), 657–674.

Riemann, B. (2004). *Collected Papers* (R. Baker, C. Christenson, & H. Orde, Trans.). From Bernard Riemanns Gesammelte Mathematische Werke, 2nd Edition, edited by Heinrich Weber, Teubner, Leipzig, 1892. Heber City: Kendrik Press.

Ritchey, T. (1983) Towards a Theory of Nonlinear Social Evolution, Liber Publications, Stockholm.

Ritchey, T. (1991). Analysis and synthesis – on scientific method based on a study by Bernhard Riemann. *Systems Research, 8*(4), 21–41. Available for download as REPRINT at: http://www.swemorph.com/downloads.html (accessed 2010-12-01)

Ritchey, T. (1997). Scenario development and risk management using morphological field analysis. *Proceedings of the 5th European Conference on Information Systems (Cork: Cork Publishing Company), 3*, 1053–1059.

Ritchey, T. (1998). Fritz Zwicky, 'Morphologie' and policy analysis. *Presented at the 16th Euro Conference on Operational Analysis*. Brussels.

Ritchey, T. (2002). Modelling complex socio-technical systems using morphological analysis. *Adapted from an address to the Swedish Parliamentary IT Commission, Stockholm, December 2002*. Available online at: http://www.swemorph.com/downloads.html (accessed 2010-12-01)

Ritchey, T. (2003a). MA/Carma – advanced computer support for morphological analysis. Available online at: http://www.swemorph.com/macarma.html (accessed 2010-12-01)

Ritchey, T. (2003b). Nuclear facilities and sabotage: Using morphological analysis as a scenario and strategy development laboratory. *Presented at the 44th Annual Meeting of the Institute of Nuclear Materials management – Phoenix, Arizona, July 2003*. Available online at: http://www.swemorph.com/downloads.html (accessed 2010-12-01)

Ritchey, T. (2004). Strategic decision support using computerized morphological analysis. *Presented at the 9th International Command and Control Research and Technology Symposium, Copenhagen, September 2004*. Available online at: http://www.swemorph.com/downloads.html (accessed 2010-12-01)

Ritchey, T. (2005a). Wicked problems. Structuring social messes with morphological analysis. *From a lecture given at the Royal Institute of Technology, Stockholm, 2004*. Available online at: http://www.swemorph.com/downloads.html

Ritchey, T. (2005b). Futures studies using morphological analysis. *Adapted from an article for the UN University Millennium Project: Futures Research Methodology Series*. Available online at: http://www.swemorph.com/downloads.html (accessed 2010-12-01)

Ritchey, T. (2006a). Problem structuring using computer-aided morphological analysis. *Journal of the Operational Research Society*, Special Issue on Problem Structuring Methods, *57*, 792–801. Available online for JORS subscribers at: http://www.palgrave-journals.com/jors/journal/v57/n7/abs/2602177a.html (accessed 2010-12-01)

Ritchey, T. (2006b). Modelling multi-Hazard disaster reduction strategies with computer – aided morphological analysis. *Reprint from the Proceedings of the 3rd International ISCRAM Conference, Newark, NJ, May 2006*. Available online at: http://www.swemorph.com/downloads.html (accessed 2010-12-01)

Ritchey, T., Stenström, M., & Eriksson, H. (2002). Using morphological analysis to evaluate preparedness for accidents involving hazardous materials. *Proceedings of the 4th LACDE Conference, Shanghai, 2002*. Available online at: http://www.swemorph.com/downloads.html (accessed 2010-12-01)

Rittel, H. (1972). On the planning crisis: Systems analysis of the 'first and second generations'. *Bedrifts Økonomen., 8*, 390–396.

Rittel, H., & Webber, M. (1973). Dilemmas in a general theory of planning. *Policy Sciences, 4*, 155–159. Amsterdam: Elsevier Scientific Publishing Company.

Rosenhead, J. (Ed.). (1989). *Rational analysis for a problematic world: Problem structuring methods for complexity, uncertainty and conflict*. Chichester: Wiley.

Rosenhead, J. (1996). What's the problem? An introduction to problem structuring methods. *Interfaces, 26*(6), 117–131.

Rush, M. G. (2010). Glossary of terms provided to the facilitation community. Available online at: http://www.mgrush.com/content/view/70/33/ (accessed 2010.08.07)

Saaty, T. (2001). *Decision making for leaders: The analytic hierarchy process for decisions in a complex world*. Pittsburgh: RWS Publications.

Stenström, M., & Ritchey, T. (2004). Scenario and strategy laboratories for an extended producer responsibility system. Available online at: http://www.swemorph.com/pdf/epr9.pdf (accessed 2009.09.12)

Taylor, T. (1967). Preliminary survey on non-national nuclear threats. *Stanford Research Institute Technical Note SSC-TN-5205-83*, Sep. 17, 1967.

Thompson, D. W. (1917). *On growth and form*. Cambridge: Cambridge University Press.

Weber, M. (1949). *The methodology of the social sciences (translated and edited by Shils EA and Finch HA)*. New York: The Free Press.

Weinberg, D. H. (1996). Galaxy structure, dark matter, and galaxy formation. *Ohio State University, Department of Astronomy*. Available at: http://arxiv.org/PS_cache/astro-ph/pdf/9610/9610003v1.pdf (accessed 2010.12.02)

Wilkinson, M. (2004). *The secrets of facilitation*. San Francisco: Jossey-Bass.

Wright, R. (2009). *The evolution of god*. New York: Little, Brown and Company.

Zwicky, F. (1947). Morphology and nomenclature of jet engines. *Aeronautical Engineering Review, 6*(6), 49–50.

Zwicky, F. (1948a). Morphological astronomy. *Observatory, 68*(845), 121–143.

Zwicky, F. (1948b). The morphological method of analysis and construction. In *Courant. Anniversary Volume* (pp. 461–470). New York: Intersciences Publishers.

Zwicky, F. (1960). A morphologist ponders the smog problem. *Engineering Science, California Institute of Technology, 24*(2), 22–26.

Zwicky, F. (1969). *Discovery, invention, research – through the morphological approach.* Toronto: The Macmillan Company.

Zwicky, F., & Wilson, A. (Eds.). (1967). *New methods of thought and procedure: Contributions to the symposium on methodologies*. Berlin: Springer.

MIX
Papier aus verantwortungsvollen Quellen
Paper from responsible sources
FSC® C105338

If you have any concerns about our products,
you can contact us on
ProductSafety@springernature.com

In case Publisher is established outside the EU,
the EU authorized representative is:
Springer Nature Customer Service Center GmbH
Europaplatz 3, 69115 Heidelberg, Germany

Printed by Libri Plureos GmbH
in Hamburg, Germany